The Incredible Experience versus Ice Age

by Rolf A. F. Witzsche

Contents

About the Illustrated Science series:
Ice Age – Climate Change

It takes an independent researcher to brake the taboos that have kept mainstream cosmology imprisoned, increasingly during the past century, even while the universal taboos are wrong. For example, Johannes Kepler broke a science taboo in the early 1600s, that had kept astronomy imprisoned to a false concept for seventeen centuries, for which epicycles and fudge factors had been invented to justify the taboos contrary to physical evidence. One can see a lot of that in astrophysics today where the evidence doesn't 'obey' the long-cherished assumptions that are nevertheless vigorously defended for numerous reasons, almost like a religious taboo. The Illustrated Science series is intended to open the scene beyond the threshold of the taboos, to where actual physical evidence speaks for itself.

The Ice Age and Climate Change sciences are riddled with assumptions that physical evidence renders obviously false. The evidence begs to be acknowledged; it promises amazing realizations when one 'listens' to what it is telling us, especially in astrophysics in connection with the coming Ice Age where so many myths abound that are simply not real, while the evidence tells us of the next New Ice Age beginning in the 2050s time frame with consequences that affect the entire world. The consequences render the areas outside the tropics essentially uninhabitable. This means that all of Canada, Europe, Russia, and parts of the USA, China, and India, need to relocate themselves into the tropics, for them to be able to exist past the 2050s timeframe. The challenge is enormous, but it can be met if the scientific imperative for it becomes understood and acknowledged.

The enormous scope of the existential challenge that the Ice Age brings with it, takes astrophysics out of the academic domain and places it into the foreground as one of the most critical issues of our time. The Climate Change effects are mere fringe effects of the changing cosmic dynamics. The big effect promises a dimmer and colder Sun with 70% less radiated energy. We can live with that by creating new agricultural infrastructures that are able to operate under such conditions. But will we do it? The task is enormous, though we have the materials and energy resources on hand. The big question therefore is; will we develop our inner resources as human beings sufficiently to get the job done? Or will we do nothing, ignore the challenge, and condemn our children and one-another to an agonizing death by starvation? That's the choice.

Towards meeting the inner challenge, I have created a series of twelve novels, the epic series, The Lodging for the Rose. And towards meeting the science challenge, I have produced several dozen exploration videos that the Illustrated Science series in book form is modeled after. Both series are the result of over thirty years of research, for which numerous elements of evidence in related fields came to light during this timeframe.

It is my hope that the enormous work that went into producing the various series will help in some degree - for humanity that we are all a part of - to write itself a ticket to have a future.

About the book itself

Do you want to know how the Galactic System operates? Here begins a journey of extraordinary discovery, powered by the near-incredible capacity of our humanity. With it we can determine the future before it happens and shape it as we wish to experience it. This is how we can master the Ice Age Challenge that is the only existential challenge that is not artificially created.

The challenge and the solutions are visible with the power of the mind. They are real. They are a scientific certainty. They are larger than all the lesser challenges and solutions in the world, combined, including the challenge to end the insanity of nuclear war, economic collapse, and the fascism of depopulation. Also, the Ice Age Challenge is a challenge that can be easily met with the mental resources that we have as human beings if we care to mobilize them, by which, when the resources are utilized, all the lesser challenges fall away by the wayside.

Life on Earth is an incredible experience. Indeed, life itself is incredible.

The human being at the pinnacle of it

And the human being at the pinnacle of it, is so complex in its design and in operation that we have barely scratched the surface in understanding it and its significance, even at the small biological level of our procreation.

Our part in the creative process

Our part in the creative process is physically so minuscule that it is barely worth the mention. We do our little part. And even this is itself impelled by impulses that we merely react to. And once our little deed is done, the rest unfolds automatically beyond our control, and largely beyond our understanding of it.

We call the offspring our children

Then, nine months later a new human being inhabits the Earth, who is eager to learn, to discover, and to develop itself. We call the offspring our children, as though we had made them, while in reality we had almost nothing to do with the incredible process by which humanity continues to exist. It would be far more correct in this light to regard all children as the children of our humanity, and ourselves likewise, and to care for the children and for one another accordingly, universally.

The child begins to discover itself

A few years down the line, the child begins to discover itself in the context of its larger world that it is a part of.

The child becomes an active participant

And in times after that, the child becomes an active participant in the dynamics of human living, with all its wonders, in as much its understanding will reach.

Humanity begins to understand its history

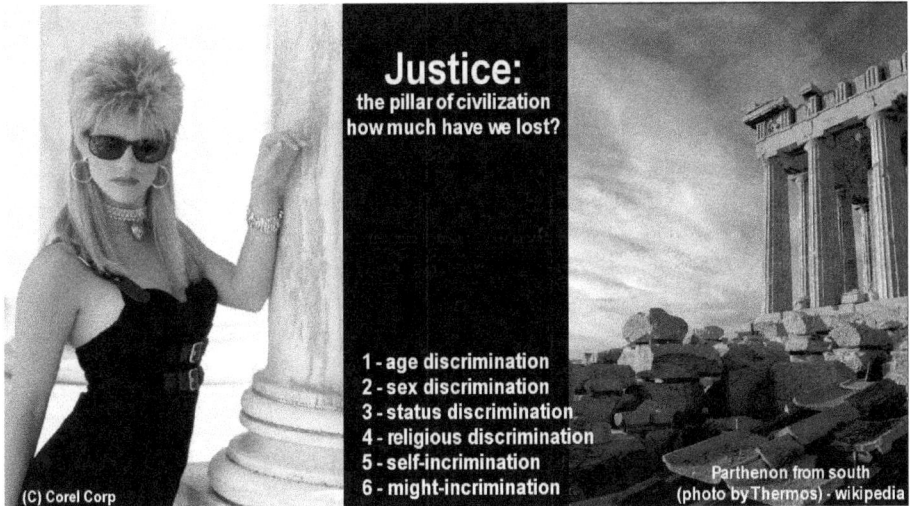

Justice:
the pillar of civilization
how much have we lost?

1 - age discrimination
2 - sex discrimination
3 - status discrimination
4 - religious discrimination
5 - self-incrimination
6 - might-incrimination

(C) Corel Corp

Parthenon from south
(photo by Thermos) - wikipedia

As grown-up adults, humanity begins to understand its history, and with it, it begins to open its eyes towards its future, both its individual future, and to some degree the future of humanity as a whole. This happens as society begins to open its inner eyes to the unseen dynamics of the world, which only science can bring to light. Here, a vista unfolds of what we have become.

The human species has developed

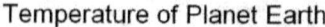

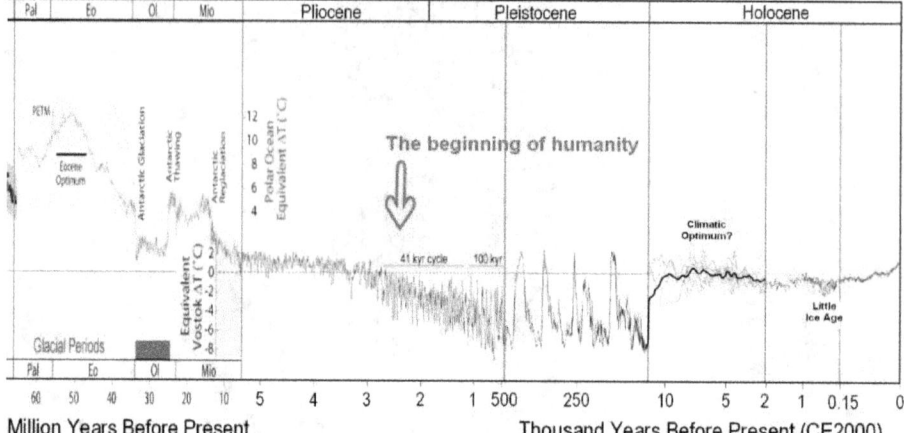

The human species has developed from its distant beginning almost entirely during the Pleistocene Epoch in geologic history that spans roughly 2 million years.

The epoch of the modern ice ages

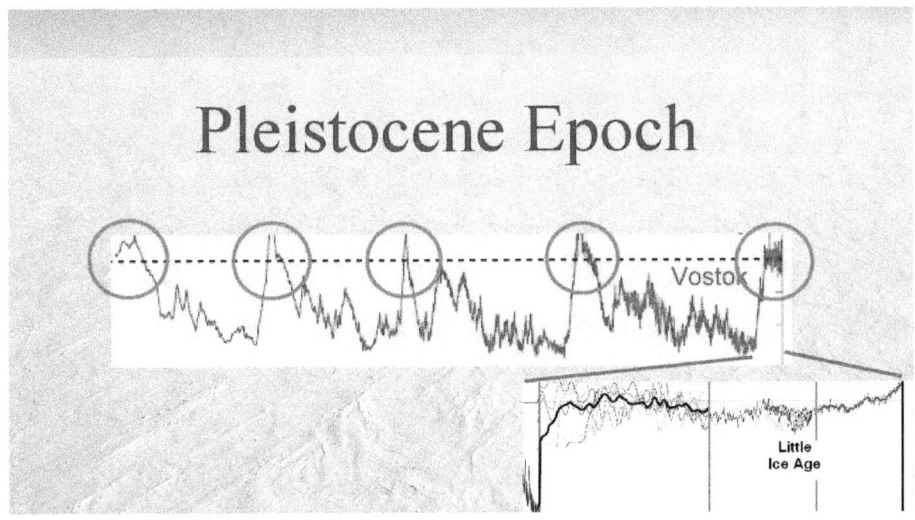

The Pleistocene is also the epoch of the modern ice ages. Less than a quarter of it is shown here. Throughout this epoch, for 85% of the time, glaciation conditions had gripped the Earth, which were interrupted only briefly with warm climates of short periods roughly 12,000 years in duration, occurring in roughly 120,000-year intervals.

What we call 'history' in modern terms, is the most recent part of the human journey that has occurred almost entirely during the current interglacial period that followed in the wake of the last Ice Age. This brief holiday from the cold, in which everything that we term civilization was developed, which has spanned slightly over 12,000 years to date, is represented by the expanded view. All that we have become - what we are today - has been developed in this relatively short period. Half-way through the period, the now common, written languages, were developed.

Later, in the period, for which the timeframe is shown expanded again, around the Little Ice Age of the Maunder Minimum, many of the great developments in science began.

In this period of the Little Ice Age

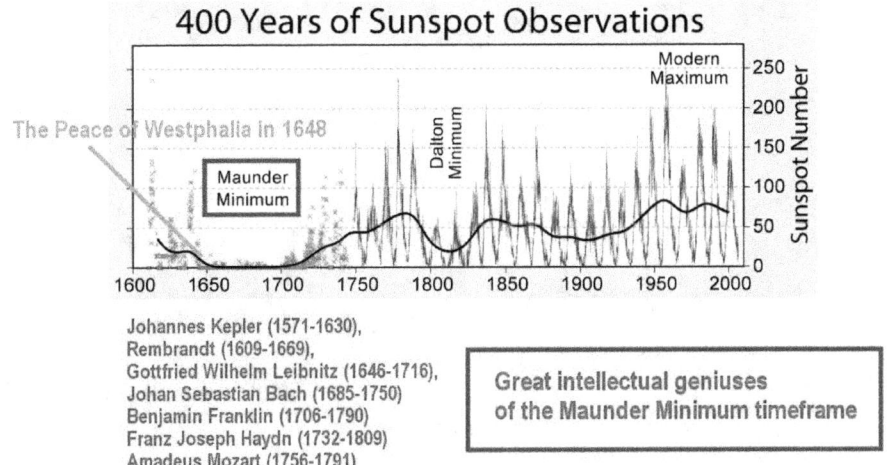

400 Years of Sunspot Observations

Johannes Kepler (1571-1630),
Rembrandt (1609-1669),
Gottfried Wilhelm Leibnitz (1646-1716),
Johan Sebastian Bach (1685-1750)
Benjamin Franklin (1706-1790)
Franz Joseph Haydn (1732-1809)
Amadeus Mozart (1756-1791)

Great intellectual geniuses
of the Maunder Minimum timeframe

In this period of the Little Ice Age, around the Maunder Minimum in solar activity, some of the great scientific geniuses had their day, whose contributions have uplifted the face of civilization. There we find Johannes Kepler uplifting science; and Bach, Mozart, and Beethoven, uplifting music; and so on.

The Little Ice Age itself

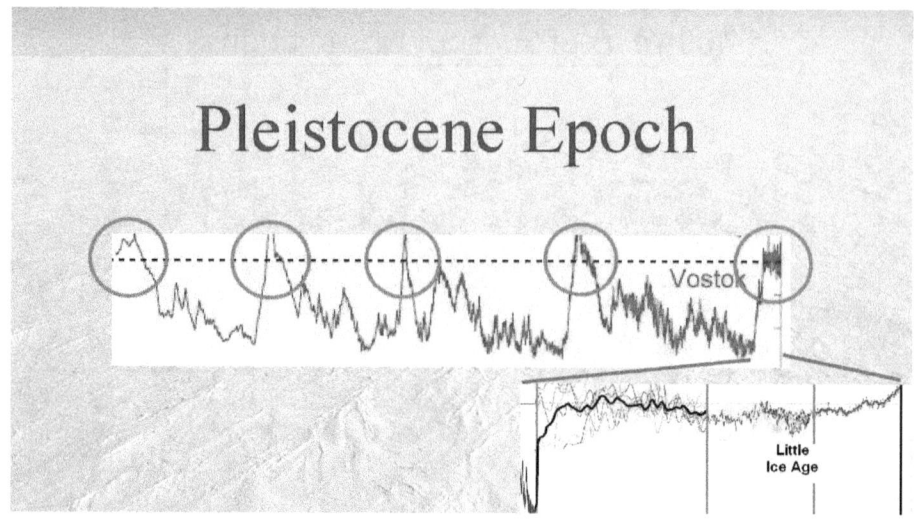

The Little Ice Age itself was but a ripple on the larger climate horizon, as was the recovery of the Sun that broke us out of the Little Ice Age and gave the world one last taste of the congenial climates in which humanity grew up.

This most recent period, the period of the global warming, that got us out of the Little Ice Age, is shown expanded again, for its more immediate importance. In this last part we find the dawn of the industrial age and the age of large-scale energy utilization, the age of technologies, air transportation, giant cities, and advances in agriculture, which together enabled our minuscule human species to increase its presence on the Earth a thousand-fold beyond what the natural world had once been able to support with its rather meagre natural resources.

The big breakout in the power of human living began in the 1800s. It started with the recovery of the Sun and lasted till around the year 2000 when the modern climate reversal began.

All that we have ever known, is now fast ending

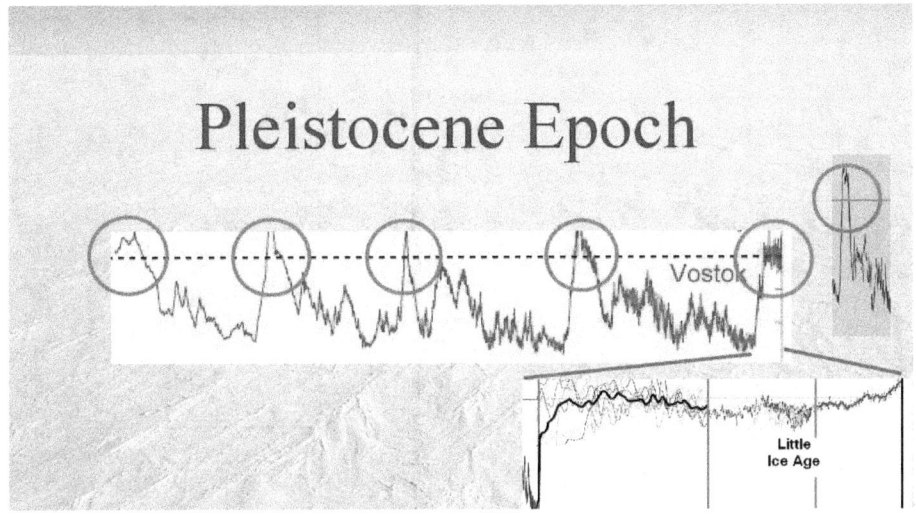

The big open question in this is, whether the solar reversal after the year 2000 can be seen as the beginning of the end of the interglacial period? The interglacial periods typically begin sharply and end sharply. That's what the ice core records convey to us. They tell us that our current interglacial was a wonderful holiday, under a brilliant Sun that is about to end, that we've taken for granted, and still do so, because this is all we have ever known and experienced. This means that the 15% portion - the slim interglacial spike that is all that we have ever known, is now fast ending for reasons that it is a climate anomaly in the over-all context, and is not representative of what is normal.

It is hard for us to imagine

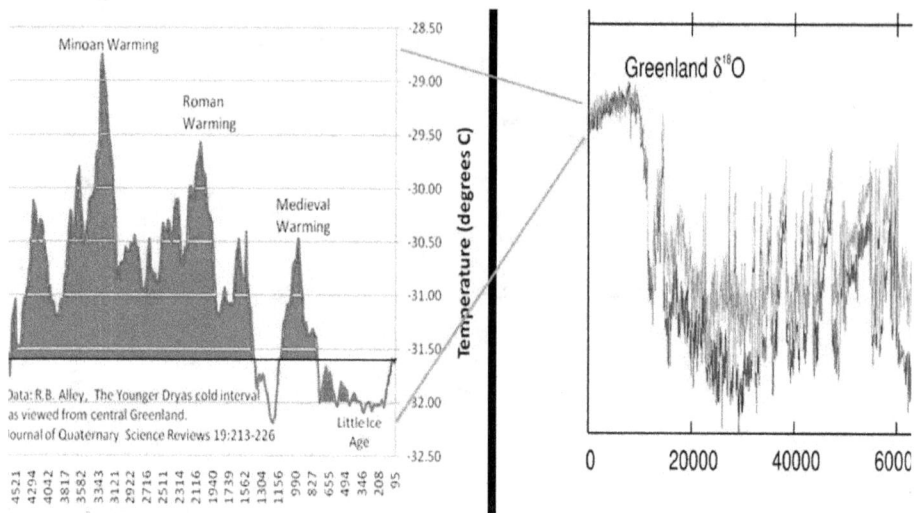

The problem is, that it is hard for us to imagine that what we have believed to be the normal world is as fleeting as a puff of smoke blown into the wind, after which the actually normal world resumes that ice core measurements tell us may be 40 times colder than the Little Ice Age had been.

Totally unprepared for living in an Ice Age world

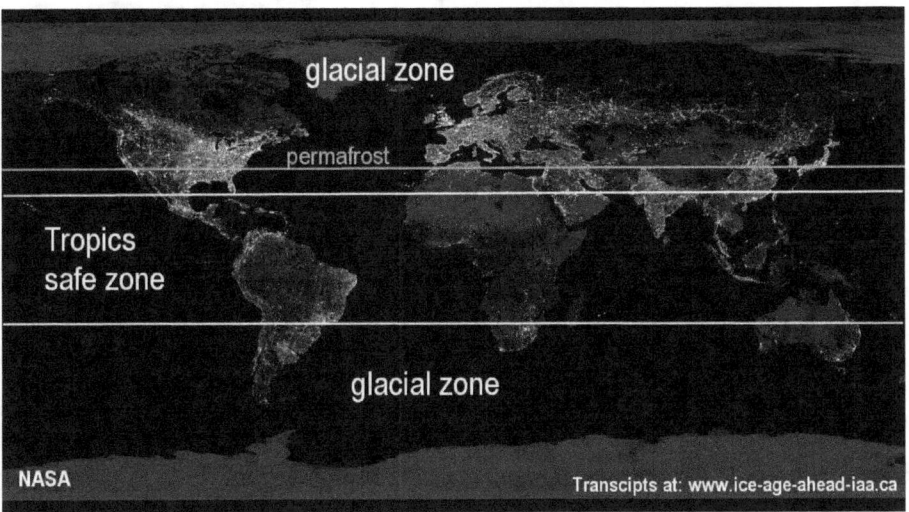

Humanity is totally unprepared for living in an Ice Age world. Most people live outside the tropics above the 23-degree latitude. These higher-latitude regions promise to become rapidly uninhabitable, possibly in less than a single year. The blue line is the permafrost line.

Can humanity survive when the Ice Age phase shift begins?

We have built our world around the basis of the present warm climate. We have established great cities and cultures in these areas where soon no one will be able to live anymore, much less grow food there. How then can humanity survive when the Ice Age phase shift begins?

The physical survival of humanity is simple

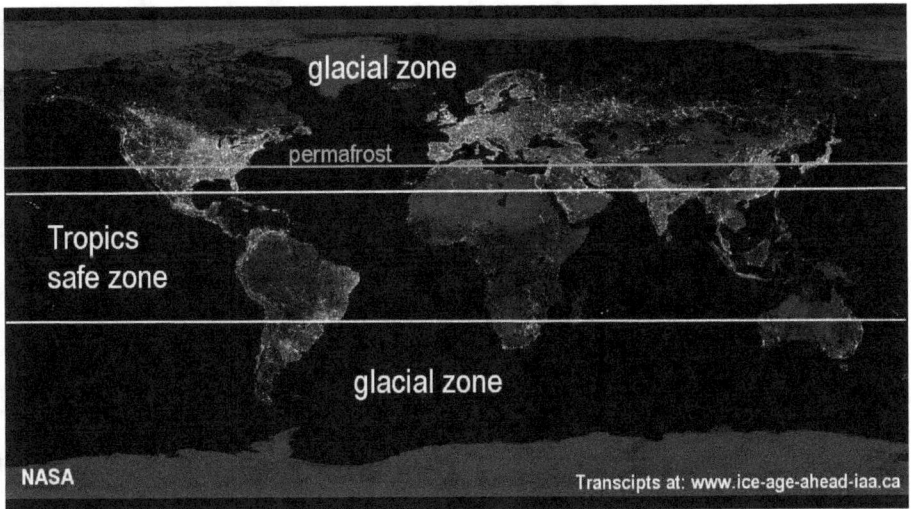

The physical option for the survival of humanity is simple. It is as simple as all the nations that live outside the tropics to relocate themselves into the tropics, which means building 6000 new cities for a million people each, complete with new industries and new agriculture.

Infrastructures placed afloat onto the sea

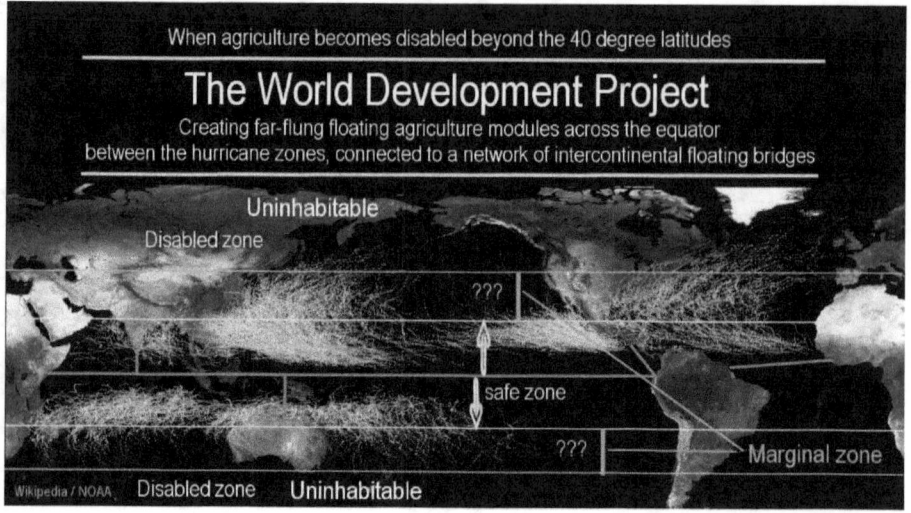

This means that most of the new infrastructures, including the cities and agriculture, be placed afloat onto the sea for the lack of suitable land in the tropics. The technologies for building the new infrastructures do exist; and so do the materials and the energy resources. But do we see anyone moving on this vital front to get the job done, or even started? We see nothing moving there. That's where the big problem lies. It lies in us, in humanity as a whole. It lies in the lacking recognition of the principles that govern the dynamics of our world, and the motivations that govern ourselves in respect to what we know.

A specific type of science for solar dynamics

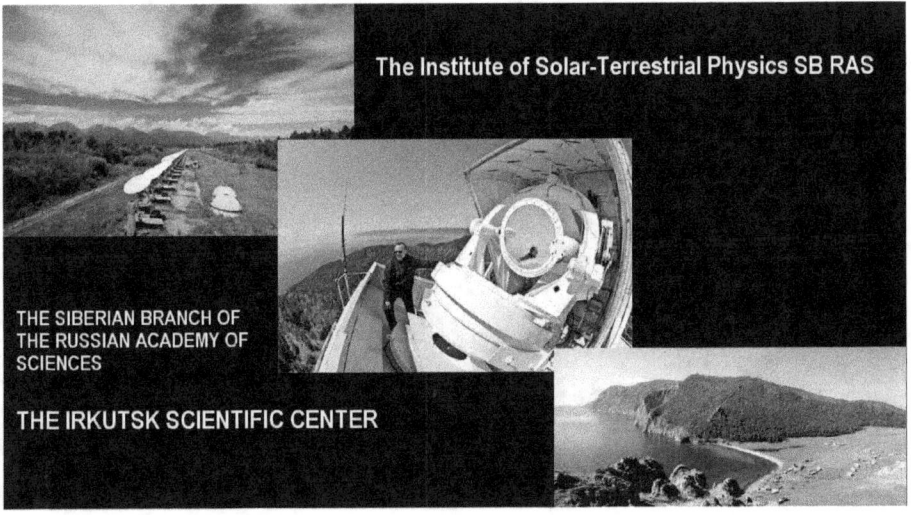

The Institute of Solar-Terrestrial Physics SB RAS

THE SIBERIAN BRANCH OF THE RUSSIAN ACADEMY OF SCIENCES

THE IRKUTSK SCIENTIFIC CENTER

Sure, we have created a specific type of science for exploring the solar dynamics. We have created the science of astrophysics for this, the queen of the sciences.

The queen of the sciences is not free

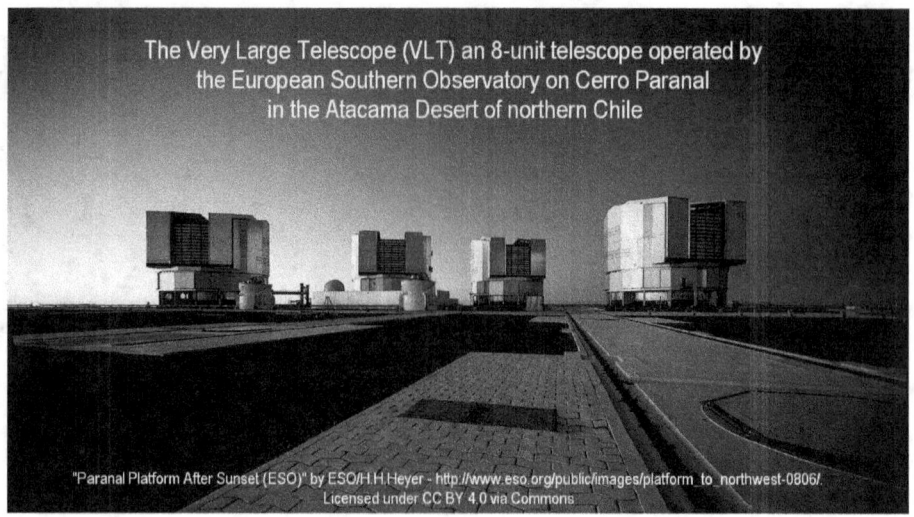

The Very Large Telescope (VLT) an 8-unit telescope operated by the European Southern Observatory on Cerro Paranal in the Atacama Desert of northern Chile

"Paranal Platform After Sunset (ESO)" by ESO/H.H.Heyer - http://www.eso.org/public/images/platform_to_northwest-0806/. Licensed under CC BY 4.0 via Commons

Unfortunately, as it is so often the case, the queen of the sciences is not free, nor has it been free for most of its history.

Astronomy had its course dictated by religion

Ptolemy
being guided by philosphy

Margarita Philosophica by Gregor Reisch, published in 1508

For a long period since the days of Ptolemy, the science of astronomy had its course dictated by religion. It was said in historic times that all the orbits in the heavens must follow the path of perfect circles, because, with the heavens being perfect, and the circle being a perfect geometric construct, no other options is theologically possible.

However, because the observed evidence didn't match the doctrine, fudge factors needed to be invented to bring the evidence into line with the doctrine. Astronomy is still stuck in this trap.

Astronomy became self-imprisoned for 1700 years

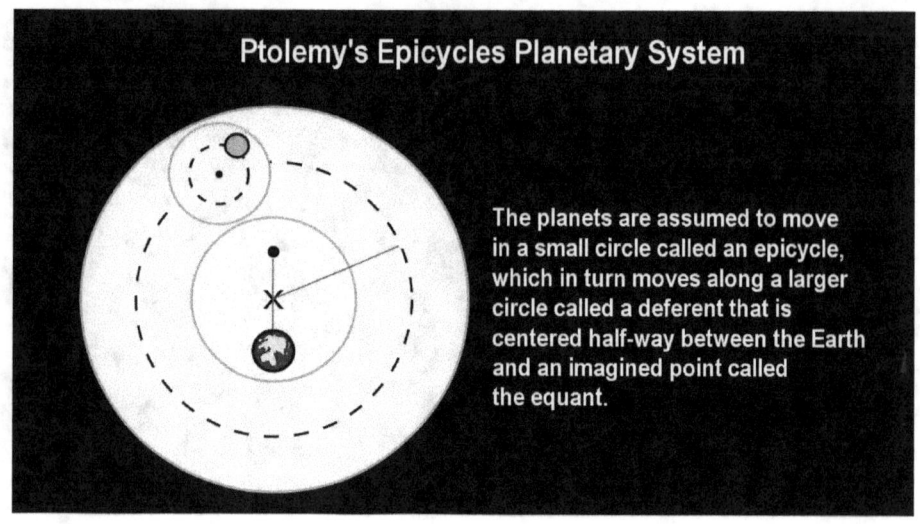

In Ptolemy's time the invented fudge factors where the imagined epicycles and the imagined equant. The inventions for self-deception were so well done that astronomy became self-imprisoned by them for 1700 years.

The epicycles had kept astronomy chained

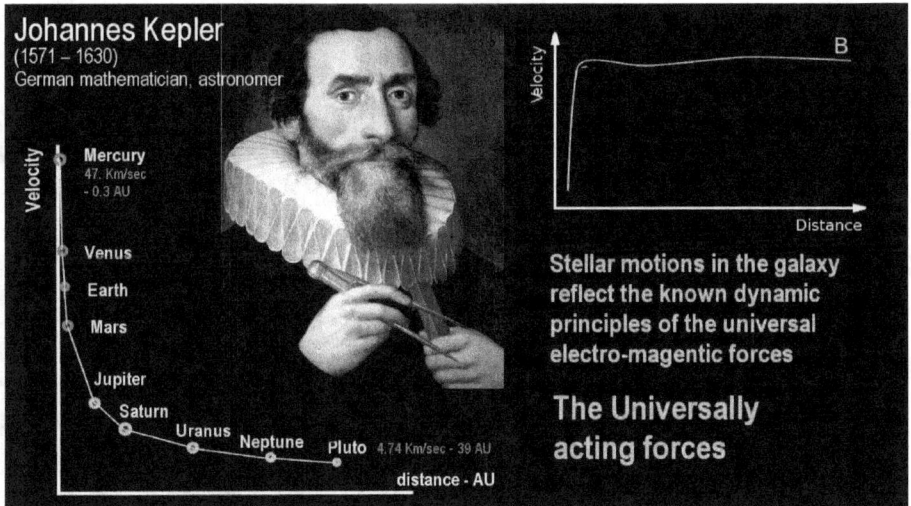

The epicycles and the equant had kept astronomy chained to a false concept for all this time until the astronomer Johannes Kepler stepped away from the doctrine, looked at the evidence, and subsequently discovered in the process of honest looking at the relatively obvious, the orbital dynamics that reflect the physical principles of mass, velocity, and gravitational interaction.

With Kepler's discovery, by simply looking at the evidence that tells its own story, a new freedom opened up in astronomy.

Nevertheless, modern astronomy is still trapped into the mysterium of imposed doctrines, such as the doctrine of orbiting stars that is completely impossible under the orbital laws that Kepler had discovered.

Black hole, dark matter, and dark energy

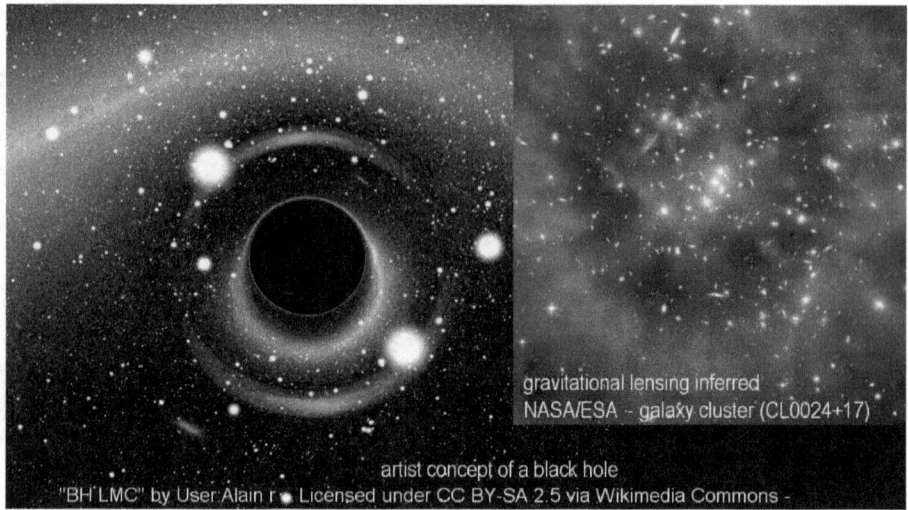

gravitational lensing inferred
NASA/ESA - galaxy cluster (CL0024+17)

artist concept of a black hole
"BH LMC" by User:Alain r Licensed under CC BY-SA 2.5 via Wikimedia Commons -

In order to make the impossible seem plausible, new fudge factors have been invented, such as the super massive black hole, dark matter, and dark energy, and so forth, which are once again, as of old, imagined phenomena that no one has ever seen, but are simply made up or inferred. Thus, astronomy still remains imprisoned under doctrines, by its own self-deception, instead of it looking for what actually moves the universe.

Doctrines keep scientific perception in chains

Entropic Energy Systems

Attribution: I, PHGCOM   wikipedia

This simply means that the dark age hasn't ended in which doctrines keep scientific perception in chains. Another imprisoning doctrine is called entropy. Entropy is a concept by which every dynamic expression in the universe is deemed to diminish towards a zero-energy state, like a wind-up toy or clock comes to a stop when the energy that has been wound up into its spring, or supplied in batteries, becomes exhausted. While the concept of entropy accurately defines these types of wind-up system, it clearly doesn't define the universe, which everyone can easily recognize.

The doctrine of universal entropy

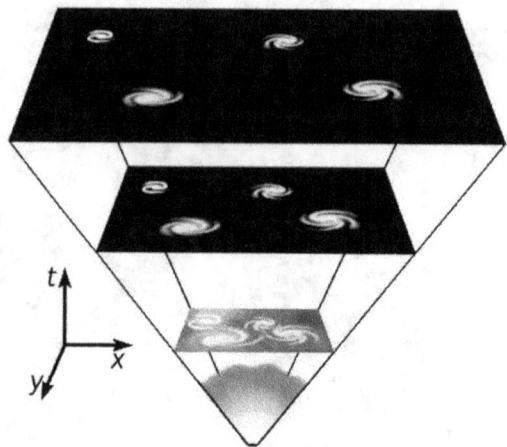

The doctrine of universal entropy has been promoted with the invention of the Big Bang theory, as a fudge factor that is designed to make the impossible seem plausible.

All that the universe will ever have

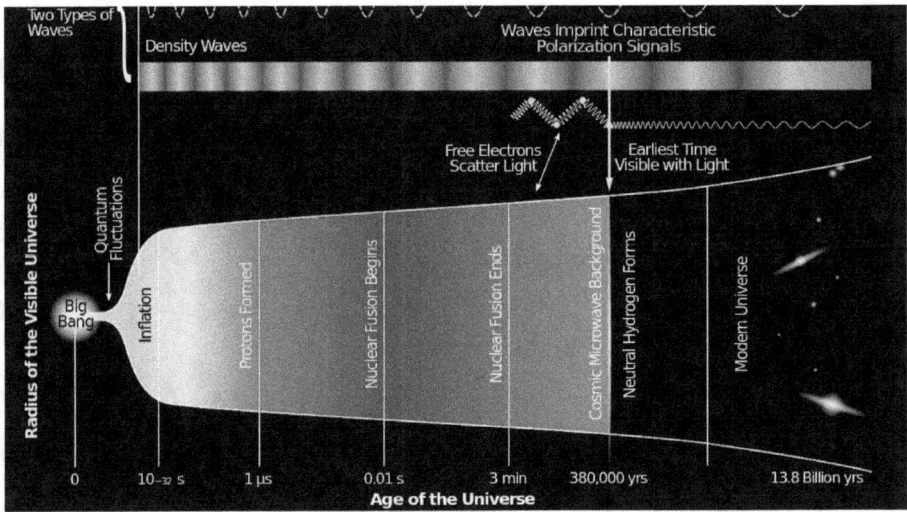

The invented theory envisions a universe that was created in a primordial, bang, in which all the energy and matter was created with which the universe is formed, and which is all that the universe will ever have, so that, necessarily, the universe is on a path of winding down towards its ultimate 'death.'

A component of the Big Bang theory

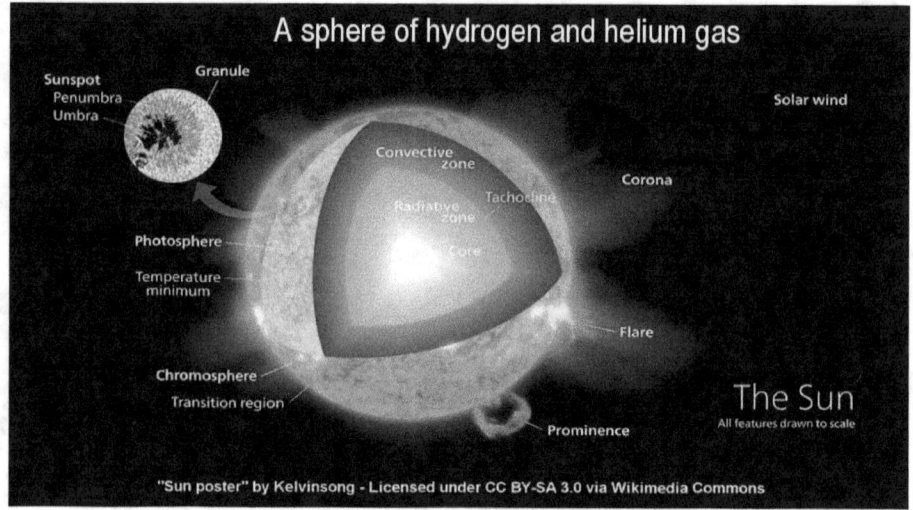

A sphere of hydrogen and helium gas

Sunspot
Penumbra
Umbra

Granule

Solar wind

Convective
zone

Corona

Tachocline

Radiative
zone

Photosphere

Core

Temperature
minimum

Flare

Chromosphere

Transition region

The Sun
All features drawn to scale

Prominence

"Sun poster" by Kelvinsong - Licensed under CC BY-SA 3.0 via Wikimedia Commons

A component of the Big Bang theory is the hydrogen model for the Sun. It envisions the Sun as a sphere of hydrogen gas that is compressed by gravity to such immense density that nuclear fusion occurs that in which hydrogen atoms are forged into helium atoms, in a process in which energy is deemed to be created that percolates to the surface and lights up the Sun to a brilliant sphere in the sky.

The Sun is deemed to consume itself

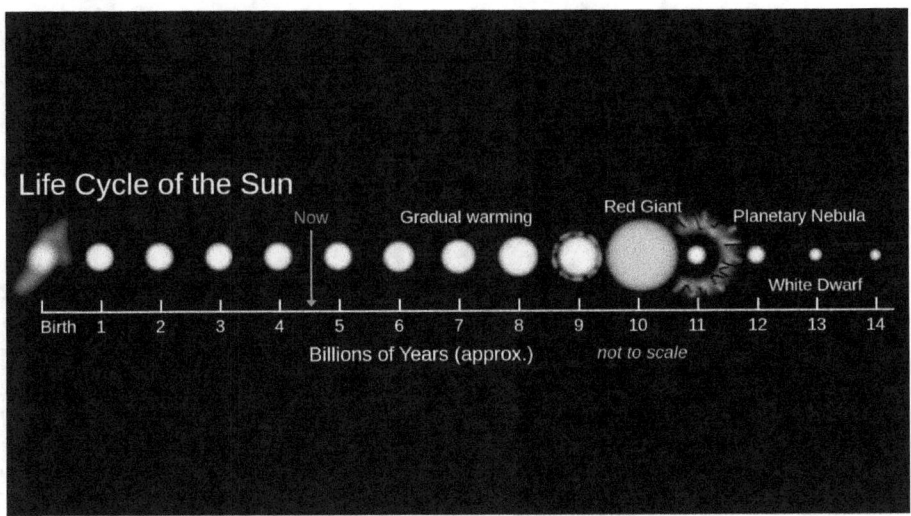

Life Cycle of the Sun

The Sun is thereby deemed to consume itself as it burns its hydrogen into helium. That's the theory. It completely adheres to the doctrine of entropy.

But the theory is not true

But the theory is not true, is it? The visible evidence doesn't support the theory. Everyone has seen the evidence bright and clear that renders the Hydrogen-Sun theory to be false, by simply looking at the sky when a rainbow appears. Seven distinct colors can be recognized in the rainbow with numerous shades of them in between, in a continuous band of color without gaps and breaks.

Continuous band of color

The same continuous band of color can be seen when the sunlight is passed through a prism. All the colors and the shades in between are contained in the sunlight, which are sorted out and separated by the prism according to their wavelength.

One sees only 5 narrow strips of light

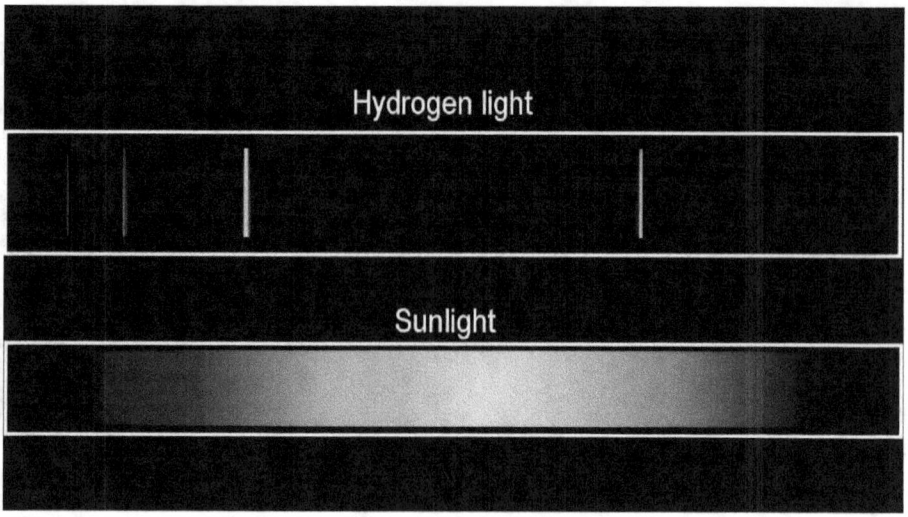

However, when one looks at the light that hydrogen gas is able to emit, in comparison with the sunlight, one sees only 5 narrow strips of light with nothing in between. Two of the strips are so faint that they wouldn't even be visible in a rainbow.

The simple evidence that one sees in comparing the sparse hydrogen-light spectrum with the actual sunlight spectrum, proves that the theory of the hydrogen Sun, is impossible, though it has become near-universally accepted as a kind of doctrine. The hydrogen Sun theory is impossible, because it cannot produce the sunlight that we see. The evidence is so dramatic that, surely, anyone would agree that the hydrogen Sun theory is false

Can you recognize that it is false?

"Can you recognize that it is false?" one might ask a schoolgirl.
"Of course, it is false! How could it not be false?" the girl might
answer.

The nature of the hydrogen atom is too primitive

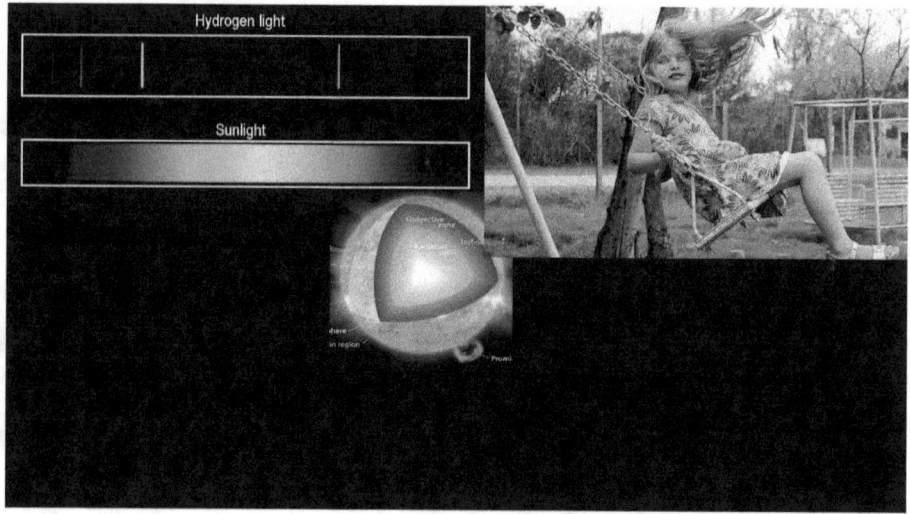

"The difference between the light that the theory gives us, and the light that the Sun gives us, is like night and day. The theory doesn't come even close. The theory is false, because the nature of the hydrogen atom is too primitive.

The hydrogen atom has only one electron

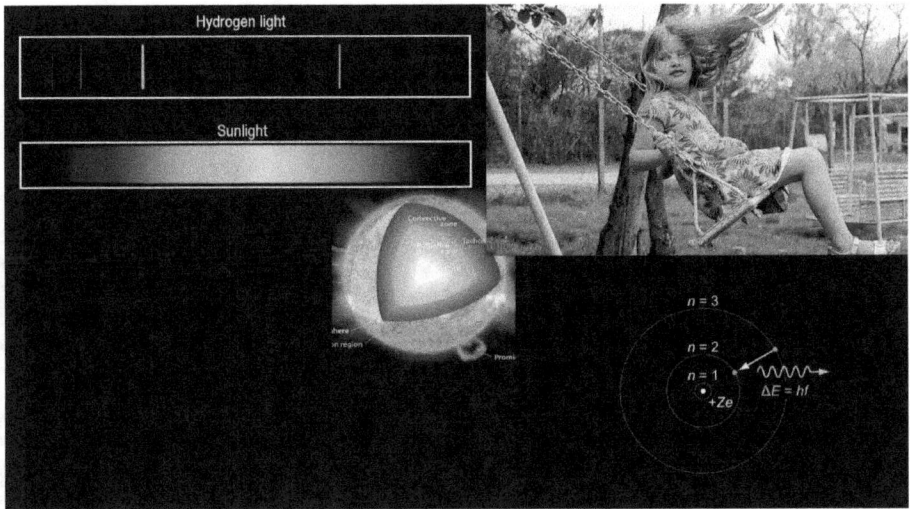

Our teacher told us that the hydrogen atom has only one electron. The teacher says that when an atom absorbs energy, its electron jumps to a higher orbit. But since it can't stay there, it snaps back, whereby the energy is re-transmitted, and flows away from it as a photon of light. With hydrogen having only one electron, the teacher says, it can produce only a few types of photons, because there exist only so many higher orbital spaces that the electron jump into, and back from. This is why only 5 different colors can be emitted by a hydrogen atom, in narrow strips.

Combine the light from a few different types of atoms

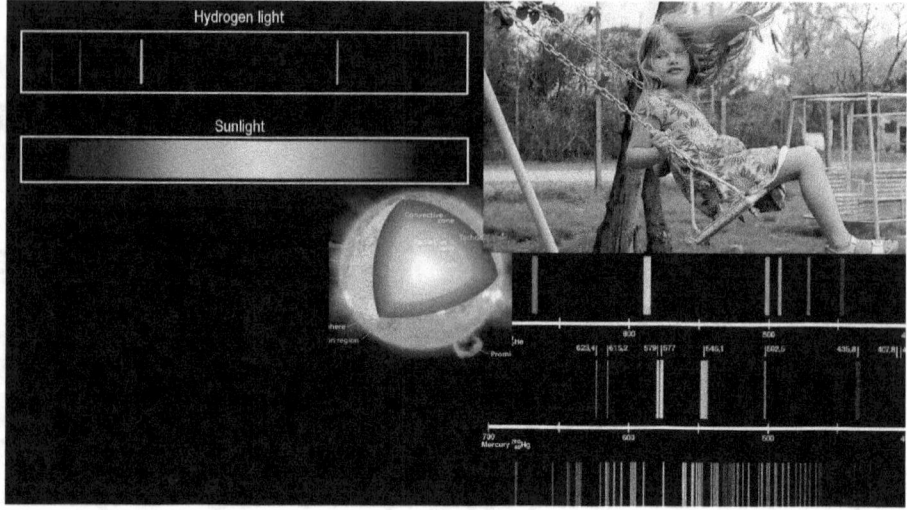

But if one was to combine the light that is emitted from a few different types of atoms, which all have different emission spectra, which in some cases also have many more of them, as in the case of uranium that has an extremely complex atomic structure, as the teacher tells us, then the result would add up to something much closer to the full spectrum of the sunlight, though it still would not be quite enough.

How about 94 of them, all emitting light together

"Open your eyes further," Johannes Kepler would say to the child if he was alive today. "Combine in your mind not only the light spectra of just four or five atomic elements. How about 94 of them, all emitting light together, each with its own array of spectra. Wouldn't their combination be sufficient to produce the sunlight that we see?

The girl would nod in agreement. "In this case, all of the atomic elements would have to be energized together, right at the surface of the Sun. But why would they be there? Wouldn't they all fall into the Sun and vanish from sight?"

"They would be there, all of them together, if they were dynamically created there, right at the surface of the Sun," Kepler would answer. "All atoms in the universe are assemblies of electrons and protons. It takes a lot of energy to combine them into atoms. But in space they exist in unbound form. In this form they are termed plasma. Also, they all have an electric charge. In an atom the charges are balanced, and thereby neutralized. In space, the charges are dynamically mixed. Your teacher may have told you about that.

The Sun is evidently not a sphere of hydrogen gas

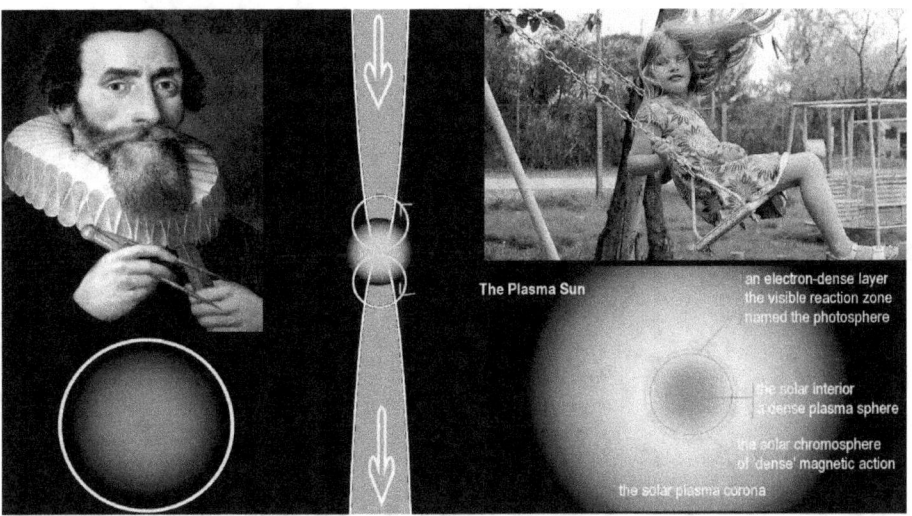

The Plasma Sun

an electron-dense layer
the visible reaction zone
named the photosphere

the solar interior
a dense plasma sphere

the solar chromosphere
of 'dense' magnetic action

the solar plasma corona

"This wasn't known in my time," Kepler would continue, "but it is known now. It is known that almost all of the mass in the universe exists in the form of plasma, at least 99.999% of it does so. Now, while the Sun is evidently not a sphere of hydrogen gas, it is possible for the Sun to be a sphere of plasma that attracts plasma from surrounding space. In fact, plasma is focused onto the Sun in highly concentrated form. This plasma is literally forced onto it. It binds it up into bundles with the electric force, and creates tightly knit atomic elements with it, right at its surface, whereby the sunlight is created that thereby appears to be the Sun's surface." Kepler might ask the girl, who would likely be smiling at this point, "Can you visualize with the power of your mind, such an atom-synthesizing process happening right at the surface on the Sun? If so, how would you know with certainty that what you see is true?"

I can see the truth with my own eyes

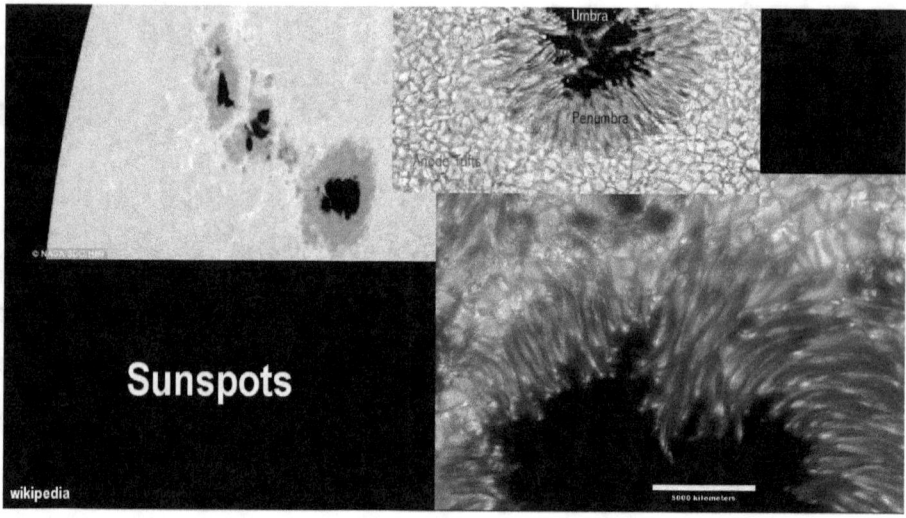

"I know it to be true," the girl would answer, "because I can see the truth with my own eyes. When one looks at the Sun through the umbra of the sunspots that are holes in its shiny surface, the Sun appears dark inside."

No one has ever seen anything different

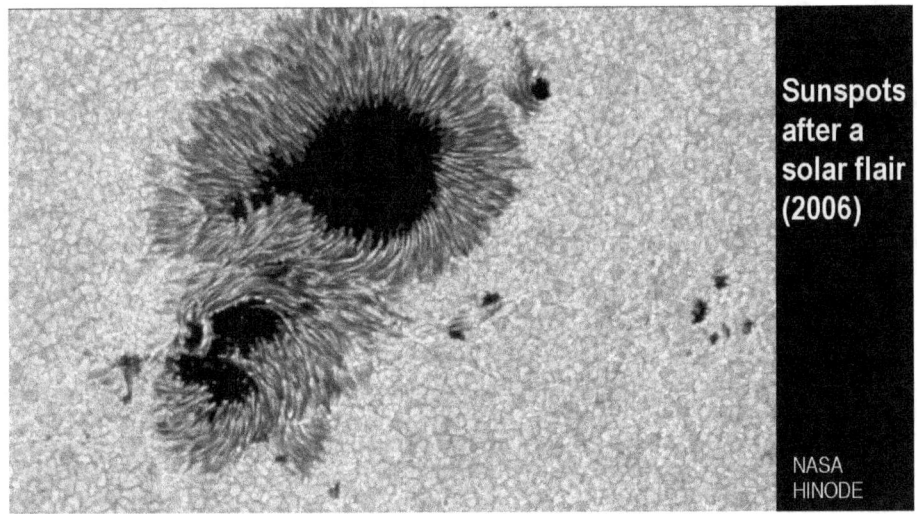

Sunspots after a solar flair (2006)

NASA HINODE

"It's plain for all to see," she would say. "No one has ever seen anything different. I asked the teacher about it. He says that this subject goes beyond the curriculum that he is allowed to teach."

I was luckier in my time

The Thirty Years' War

The miseries of war; No. 11, "The Hanging"
by Jacques Callot (1592–1635)

"I was luckier in my time," Kepler would answer. "In my time, during the 30-Years War, everybody was too busy killing one-another. The exotic doctrines carried little weight then, for which the epicycles and fudge factors were invented."

I was free to look at the universe honestly

Kepler's Three Laws of Planetary Motion

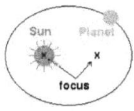

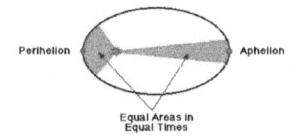

Equal Areas in
Equal Times

$$\frac{P_1^2}{P_2^2} = \frac{R_1^3}{R_2^3}$$

I. The orbits of the planets are ellipses, with the Sun at one focus of the ellipse.

II. An orbit sweeps out equal areas in equal times as the planet travels around the ellipse.

III. The ratio of the squares of the revolutionary periods for two planets is equal to the ratio of the cubes of their semimajor axes:

http://csep10.phys.utk.edu/astr161/lect/history/kepler.html

"I was free to look at the universe honestly," Kepler would say to the girl. "It is amazing what one begins to discover by becoming free to what unfolds in the perspectives of the mind. It becomes an incredible experience to experience oneself as a human being.

Grander than the grandest landscape

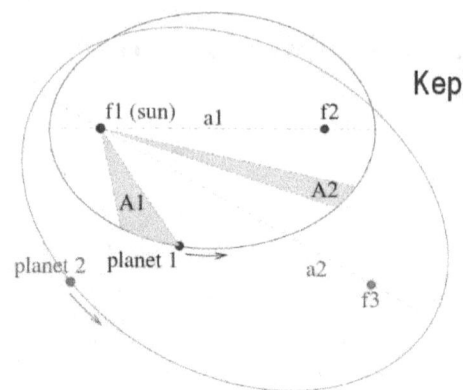

Kepler's three laws of planetary orbits

(1) The orbits are ellipses, with focal points $f1$ and $f2$ and $f3$ respectively.

(2) A1 = A2, both in area and in transit time.

(3) The total orbit times for planet 1 and planet 2 have a ratio $a1^{3/2} : a2^{3/2}$.

What comes to light there is grander than the grandest landscape and the brightest rainbow. Nothing comes even close to that, just to be able to see this.

I wrote my discoveries down in a book that I named the New Astronomy. I didn't care whether anyone would understand me for another thousand years. I knew that what I had discovered is the truth, and I was glad to share it."

Utilized to calculate spaceflight trajectories

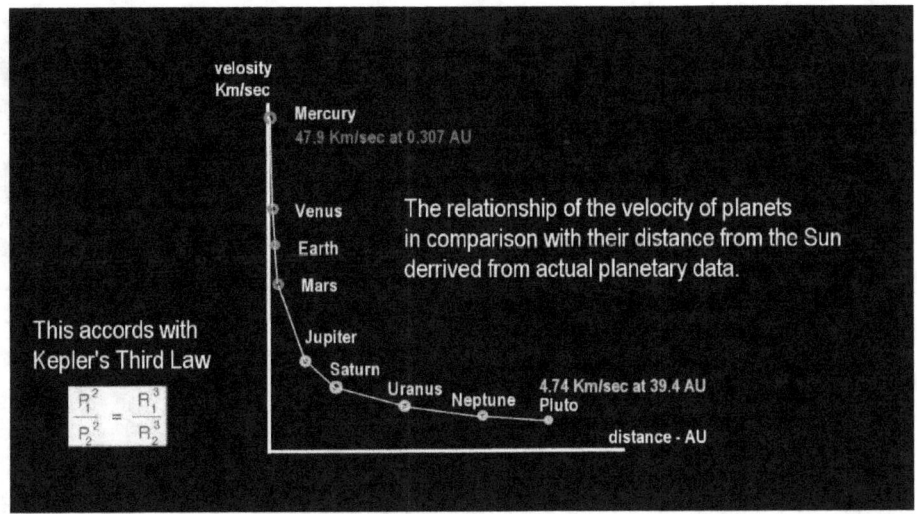

"Well, it didn't take a thousand years," the girl would answer back. "Your discoveries are now routinely utilized to calculate spaceflight trajectories. What you saw has timeless significance."

This is why you should be afraid for your life

"This is why you should be afraid for your life," Kepler would answer back. "The history books may tell you that I starved to death. I died in the fields out of weakness while searching for food for my children. Too much had been destroyed in the 30-Years War. Too many people had been killed, and farms burned to the ground. Food production had collapsed. More people likely starved to death than were killed in the war. I am saying this, because you still have a chance to avoid the same fate.

"But before you will ask me why I am saying this, let me ask you instead how secure our Sun really is, as you now see it, which is a sun that depends on energy supplied from space."

Our Sun cannot be a secure Sun

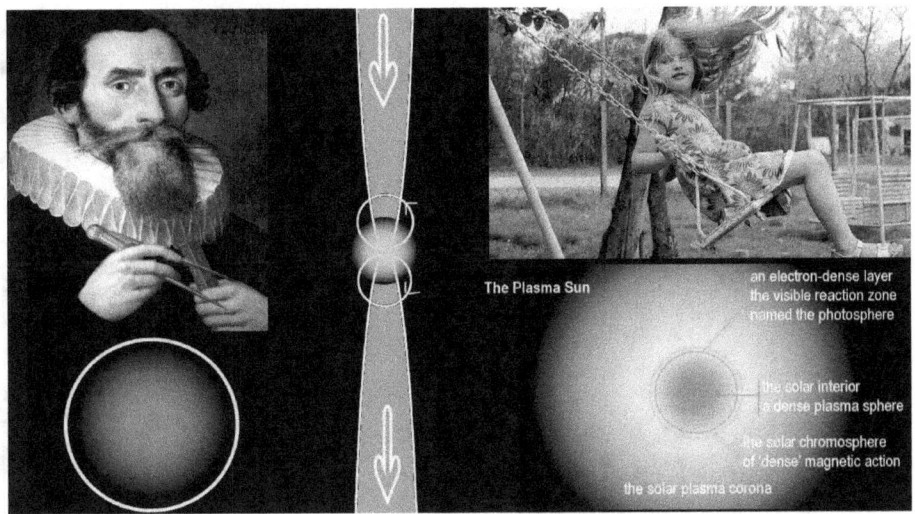

The girl would nod and answer a while later, hesitatingly, "Our Sun cannot be a secure Sun if it depends on external energy flowing to it from space. But how certain can we be that it is vulnerable?"

It is a fact in basic physics

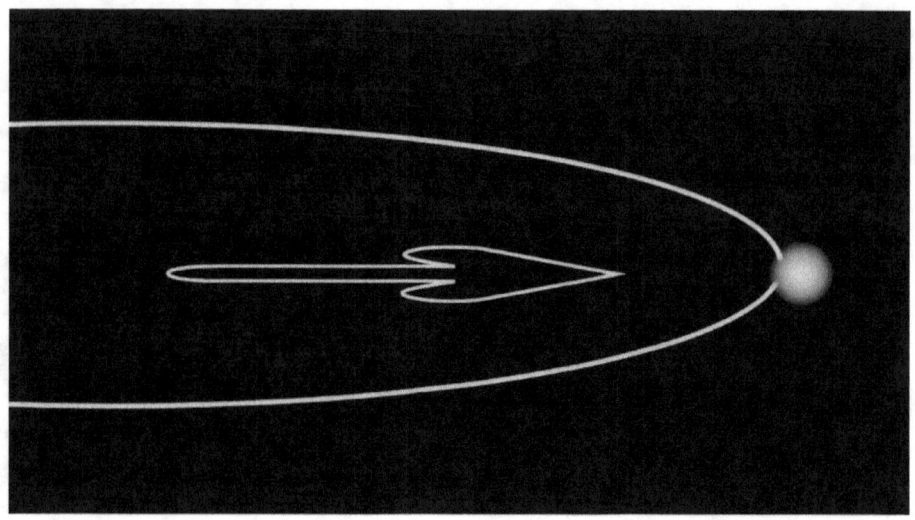

"You should be able to answer this yourself," Kepler would say. "Your teacher may have presented the answer in your physics class. It is a fact in basic physics that when an electric current flows in two parallel wires in the same direction, the wires attract one another by magnetic interaction, termed the Lorenz force. While wires are typically held in place, the flowing electric plasma in space gets pinched together by the attracting electromagnetic forces, which thereby increases the electric density and the resulting magnetic fields, which in turn increases the pinch effect still further. This goes on until the magnetic fields get all tangled up.

The flowing plasma gets curled backwards

At this point the flowing plasma gets curled backwards and collects under a magnetic dome where it becomes further concentrated. Under the resulting pressure the concentrated plasma escapes through the open hole and gets focused onto the Sun where it is consumed by the process of atomic synthesis on the surface of the Sun."

I can see what you are getting at

The girl begins to smile again. "I can see what you are getting at, she interjects.

When the interstellar plasma streams become too weak

"When the interstellar plasma streams become too weak for the magnetic fields to become tangled up by the pinch effect, so that the plasma doesn't get flipped back and becomes concentrated thereby, to be focused unto the Sun, then the plasma density around the Sun becomes too weak for the plasma-fusion to be possible on the solar surface, on the grand scale as we have it now, which lights up the Sun, so that the electromagnetic fields that focus the plasma may break down. The Sun would go inactive then, when this happens, whatever this may mean. Our familiar Sun will then no longer exist in the way we see it today. It may become dim."

No one has yet seen what our Sun will then be like

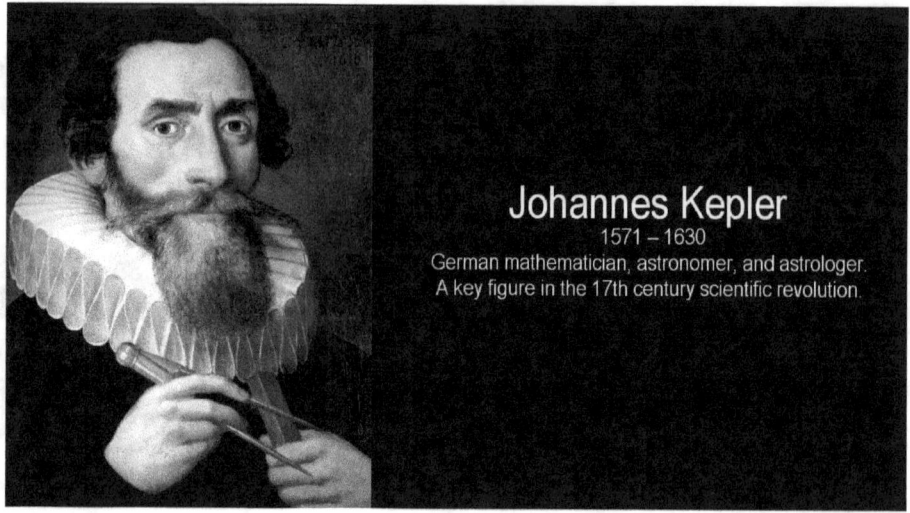

Johannes Kepler
1571 – 1630
German mathematician, astronomer, and astrologer.
A key figure in the 17th century scientific revolution.

"No one has yet seen what our Sun will then be like," Kepler might interject, "but we can know this precisely, and long before it happens, because as human beings we have the capacity to see with the mind far into the future, by understanding the principles that govern the physical universe."

Isn't this amazing what we can be aware of

"Isn't this amazing what we, as little human beings, can be aware of," the girl might say. We can see what no bird or animal can as much as dream of, but we can see it happening in our mind as clearly as if it was plainly visible and was happening right now, instead of in future ages. Wow, what an experience! That's better than riding the swings. By understanding the basic principles of the universe, we can see the future long before it happens, in all critical details. Is this really possible? It would be the most incredible experience of them all," the girl would add.

I was fortunate Kepler might add

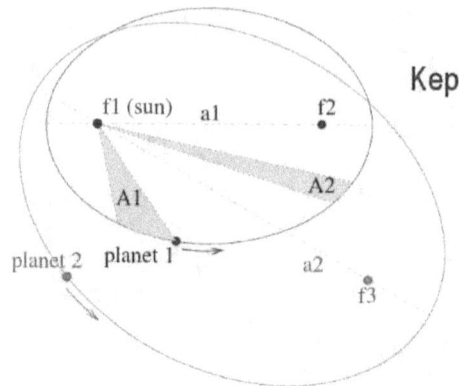

Kepler's three laws of planetary orbits

(1) The orbits are ellipses, with focal points $f1$ and $f2$ and $f3$ respectively.

(2) A1 = A2, both in area and in transit time.

(3) The total orbit times for planet 1 and planet 2 have a ratio a1^3/2 : a2^3/2.

"I was fortunate in what I did," Kepler might add, "because I had an extensive set of accurate astronomical measurements prepared for me by Tycho Brahe. I simply took his measurements, looked at them, and in the process of looking at them with the mind's eye it became evident to me what was really going on. Then, once I saw the answer that had evaded astronomers for 15 centuries, and knew that what I saw is true, I stood on the same ground that the slave boy had stood on in Plato's Mino Dialog, who had discovered the principle for doubling the area of a square, and could say in the end with absolute certainty that what he discovered is correct by simply looking at the visual evidence.

To discover the principle for doubling a square

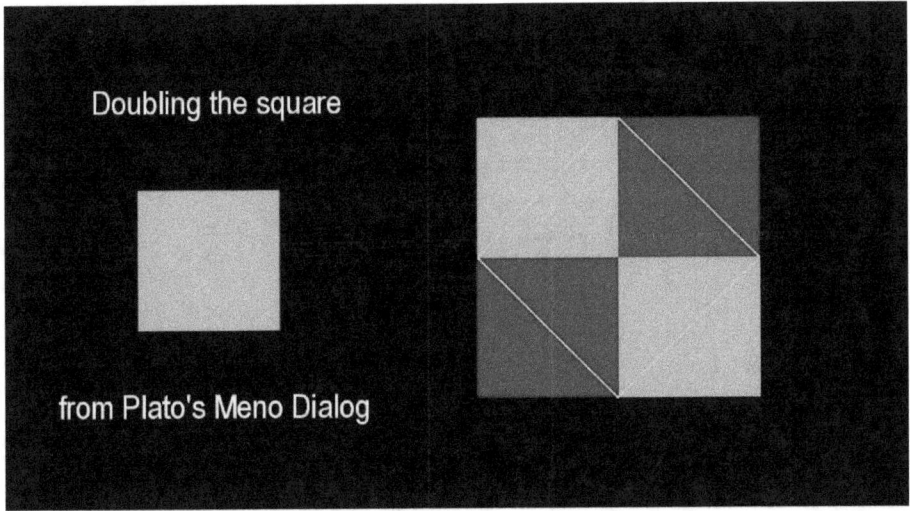

Doubling the square

from Plato's Meno Dialog

"The slave boy in the Meno dialog is guided by Socrates to discover
the principle for doubling a square. He is guided to create a new
square that is 4 times as big as the original square, and then divide
each of the resulting four squares in half along the diagonal, which
results in a square created by the diagonals that is twice as big in
area than the original square. The slave boy could look at the result
and say with absolute certainty that the resulting square is twice as
big in area as the original square, by simply counting the triangles as
evidence. He could see the evidence reflecting the principle. He
knew the result to be true."

You can experience today the same certainty

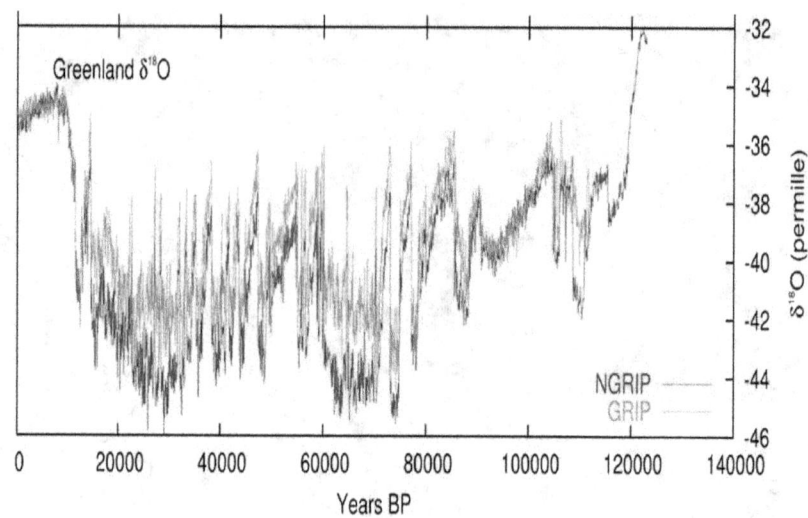

Kepler would add, saying to the girl, "You can experience today the same certainty in seeing the future before it happens, on the basis of the principles that you now know, and by seeing the reflection of these principles in measured historic physical events. You have amazing measurements available to you from the big ice coring projects that were completed at around 2003. You can see in them the principles reflected that you have come to understand as demonstrably truthful."

In the big Greenland ice cores

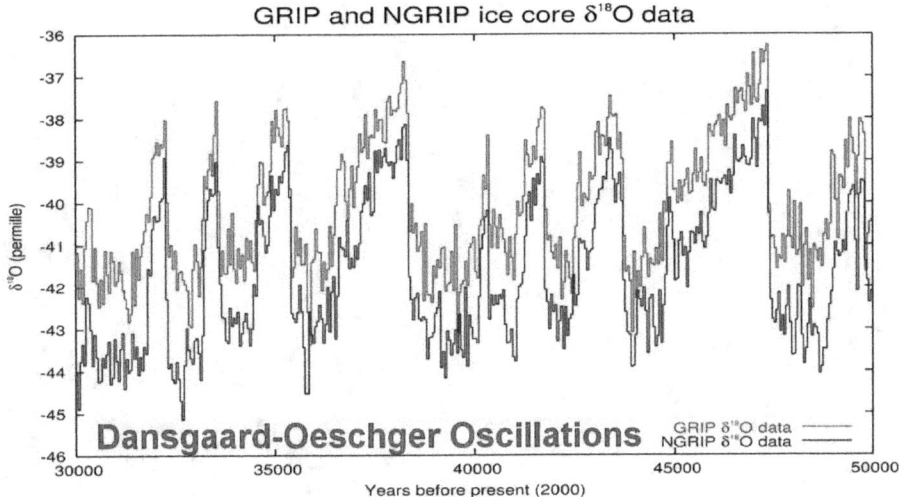

GRIP and NGRIP ice core δ¹⁸O data

"In the big Greenland ice cores you have evidence measured for you, of large, steeply rising climate oscillations from deep glacial conditions all the way to near interglacial conditions and then back again, occurring in intervals of 1470 years, called the Dansgaard Oeschger Oscillation. Some of these oscillations were up to 40 times larger than the cooling had been, that had been experienced during the Little Ice Age in the late 1600s, that had devastating effects for many people."

The coming Ice Age with an inactive Sun is real

Antarctica
by Vincent van Zeijst (wikipedia)

"Are you saying that the Sun going inactive for the next Ice Age is not just a theoretical possibility, but is going to happen with the same certainty with which the slave boy in the Meno Dialog knew that he had doubled the square," the girl might interject. "Is this what you meant when you said that I should be afraid for my life? Did you mean this as a consequence of the Sun going inactive?" Kepler would nod. "What we can see as the expression of universal principles is real. The coming Ice Age with an inactive Sun is real. The harsh consequences that result take astrophysics out of the academic domain and into the real world. Yes, my dear, you should be afraid of loosing your life, because you will starve to death with absolute certainty if the infrastructures are not built for continuing agriculture under an inactive Sun. We are on the path already. The great phase shift to the next Ice Age is already in its pre-stage. We are moving towards it. It will happen as a sudden event, possibly in the 2050s. There are resonance cycles happening in all the big plasma streams, like the resonance cycles of your swing set. We can see their effect. When the big cycles come together at their minimal points and interact, big effects happen. The timing is predictable, and the consequences are predictable too. The plasma Sun is loosing steam. NASA's Ulysses spacecraft saw the beginning of it. The plasma density is fading fast. In its inactive state the Sun will diminish and become a red star, as the inactive stars are called. In its inactive state the Sun won't support agriculture as it is known today, even in areas that are not covered in ice"

NASA's Ulysses spacecraft

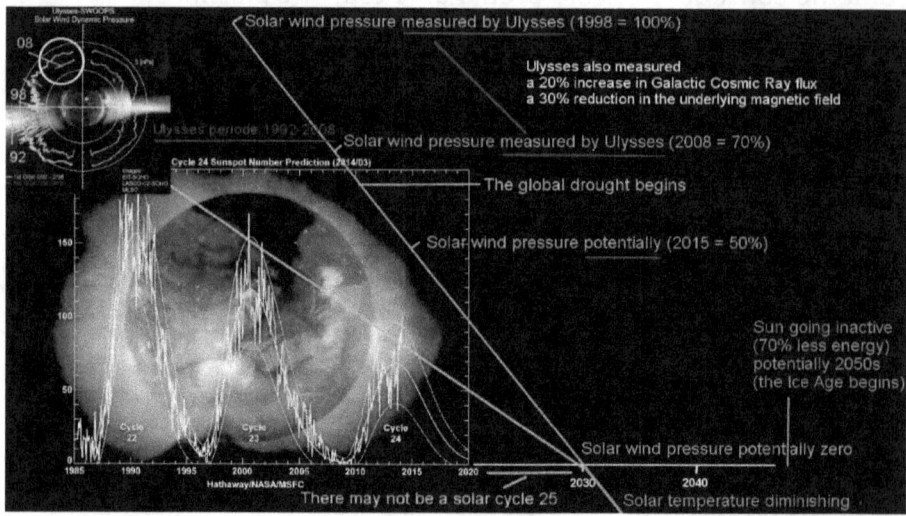

"Your NASA's Ulysses spacecraft has measured a 30% drop in solar-wind pressure in the short span of just ten years ending in 2008. That's huge," Kepler would say. "At this fast rate of fading out, the solar wind may cease to flow altogether in the 2030s. The sunspot cycles are likewise diminishing at roughly the same rate, which means that the solar activity is diminishing. The fading solar wind reflects that. This is all clear evidence that the plasma streams that are focused onto the Sun are dramatically weakening."

Does this mean that the Sun could go inactive?

Antarctica
by Vincent van Zeijst (wikipedia)

"Does this mean that the Sun could go inactive any time after the 2030s," the girl would interject. "Does anyone realize what this means?"

"That's the wrong question," Kepler might answer. "The big question is: Do you know what this means? Are you willing to make the effort to assure that you will live? Are you willing to do whatever it takes? The phase shift may not happen before the 2050s. This leaves you 30 years to get the job done, which is not much time for the nations of the world to relocate themselves into the tropics," Kepler might add. "You ask: Does anyone realize what this means? The answer is, Yes. A human being can see precisely what this means, though few people presently recognize themselves as such. You, yourself, may not see precisely what happens when the primer fields collapse, or only one of the nested primer fields that focus plasma onto the Sun. You understand the principle to some degree, and have an inkling of what happens when the principle is no longer expressed.

73

"At this point many potential events could happen. It might be that the plasma fusion on the Sun will then no longer be possible, and stop. The photosphere might vanish. In this case previously created atomic elements would no longer flow away, but fall back into the Sun. They would fall deep into the Sun until their atomic structure would become crushed. This could happen. In this case the invested binding energy would be released in nuclear fissioning. Do we see evidence for this happening? Do we see the red-dwarf stars as evidence for that? Evidence exists that three quarters of all the stars in our galaxy are presently inactive stars of the red-dwarf type? This adds up to 300 billion inactive stars. But is our Sun destined to go this route and become a red-dwarf star? Or will our Sun, when its closest primer fields become inactive, simply continue on with a lower intensity plasma fusion and a lower surface temperature of the kind we see in the red-dwarf stars? These are the questions you need to ask, and keep on asking until you know with certainty what the future of the Sun will be."

Comparison of stars - CC BY 3.0 wikipedia - ESO/M. Kornmesser

Super Giant
Star/Sun
100,000

dwarf Giant Star
1 billion

the Sun
among stars

90 billion

300 billion

Left to right: a red dwarf, the Sun, a blue dwarf, and R136a1

Kepler would have been right had he said all this. Two major types of stars have been discovered to exist in our galaxy, named the Milky Way Galaxy, which contains an estimated 400 billion stars. One of these is the inactive type, the red type. Most of this type are of the red-dwarf type.

The low surface-temperature category

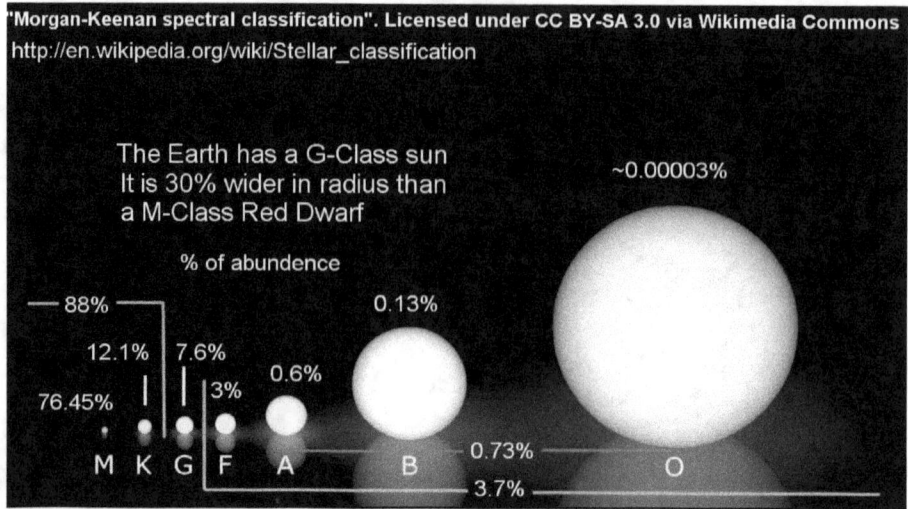

An estimated 75% of all the stars in the galaxy are of the M category, the low surface-temperature category. The red-dwarf stars dominate this category. They come in different sizes but are typically so small and so dim that they are hard to detect. Johannes Kepler would not have seen them.

The closest red dwarf in the stellar neighbourhood

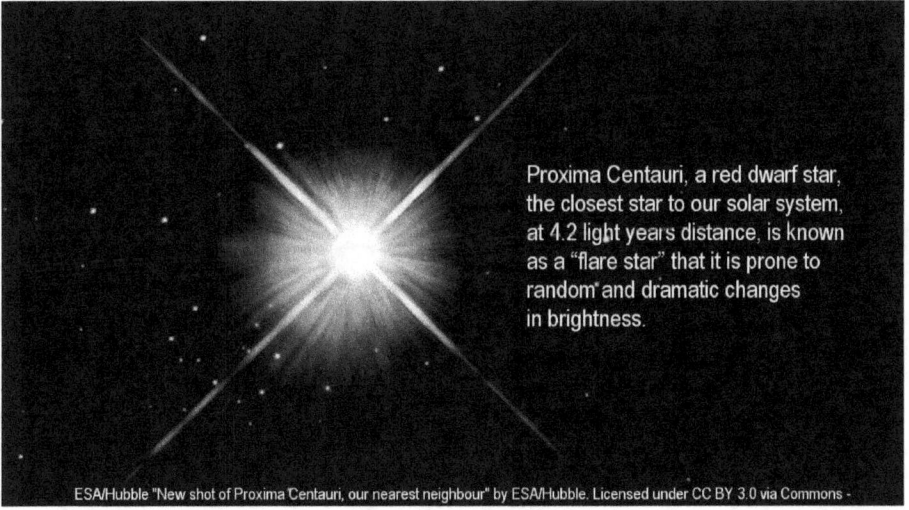

Proxima Centauri, a red dwarf star, the closest star to our solar system, at 4.2 light years distance, is known as a "flare star" that it is prone to random and dramatic changes in brightness.

ESA/Hubble "New shot of Proxima Centauri, our nearest neighbour" by ESA/Hubble. Licensed under CC BY 3.0 via Commons -

Even the Hubble space telescope can see them just barely. The image shown here is of the closest red dwarf in the stellar neighbourhood. It is a part of the nearest solar system, named the Centauri system. It is located a mere 4.2 light years distant. The number of red dwarfs in the galaxy has been estimated by what has been observed in the stellar neighbourhood.

The red dwarf is surrounded by a cloud of atomic elements

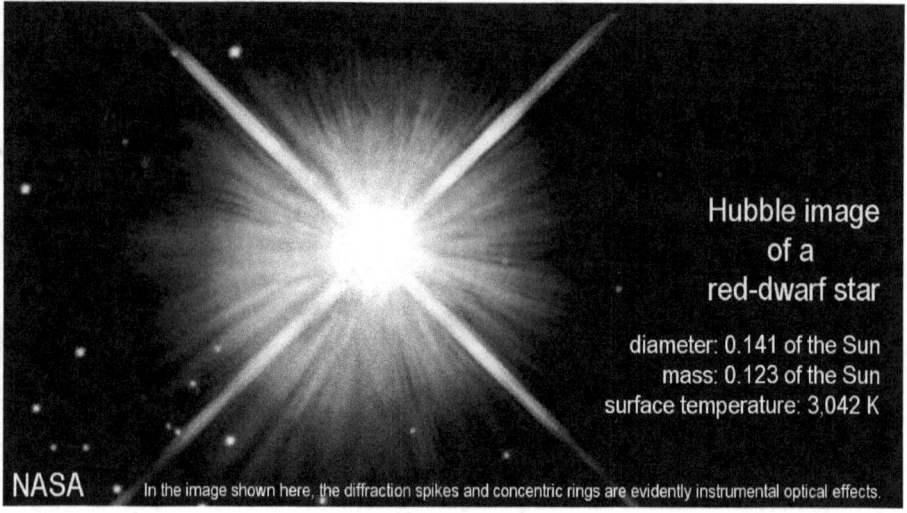

Hubble image
of a
red-dwarf star

diameter: 0.141 of the Sun
mass: 0.123 of the Sun
surface temperature: 3,042 K

NASA   •   In the image shown here, the diffraction spikes and concentric rings are evidently instrumental optical effects.

The Hubble image indicates that the red dwarf is surrounded by a cloud of atomic elements. This is what one would expect to see for an inactive star that attracts gravitationally the previously synthesized atomic elements, which no longer flow away with the solar wind. They would emit light by their interaction with the plasma of the original star. The light would increase in intensity with the increasing gravitational pressure in the plasma of the star, towards its center, until the atomic elements would fission and light up the central sphere.

The red dwarf might also be simply a small star that is powered by less intense plasma fusion with a lower surface temperature and lesser solar winds, so that the synthesized atomic elements form a haze around the star. Is this what we see here? Is this the destiny of our Sun?

Of the category M type, at 4,000 degrees Kelvin

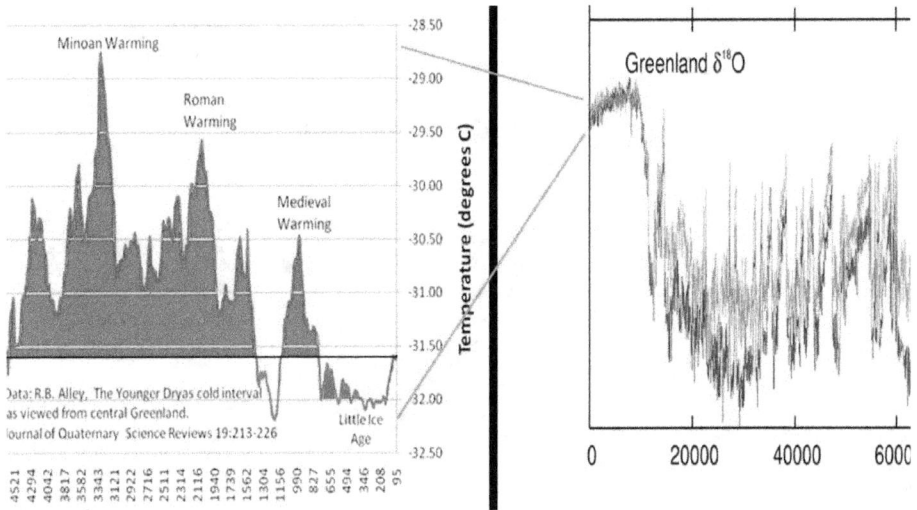

The effective surface temperature would be significantly less if our Sun would be powered by a low-level plasma-fusion process. Its surface temperature would then be of the category M type, at 4,000 degrees Kelvin, or slightly less. This would be enough to shift the climate into deep glaciation conditions that we had in the previous glaciation period that we call the Ice Age.

The sunlight spectrum would be shifted to a lower profile

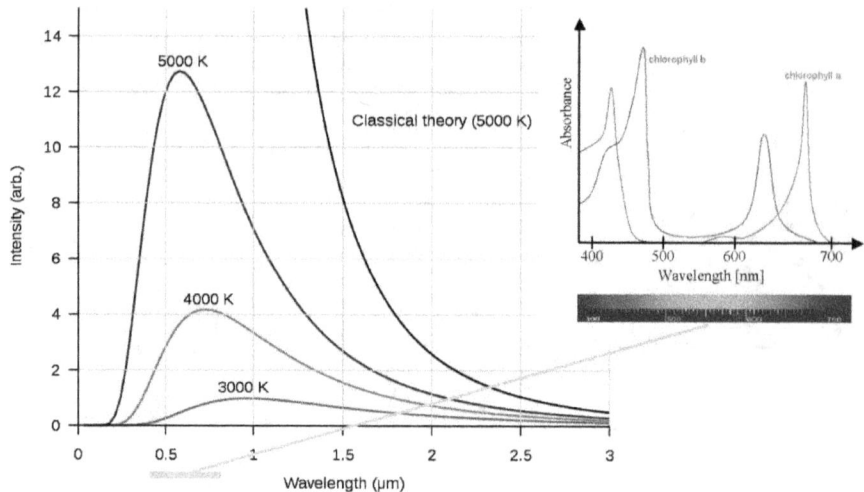

In this case the sunlight spectrum would be shifted to a lower profile but would remain in range for the chlorophyll to work. The effect will likely reduce the radiated energy that we receive on Earth to a mere 30% of what we receive today. It promises to have huge consequences.

The small size of the world population

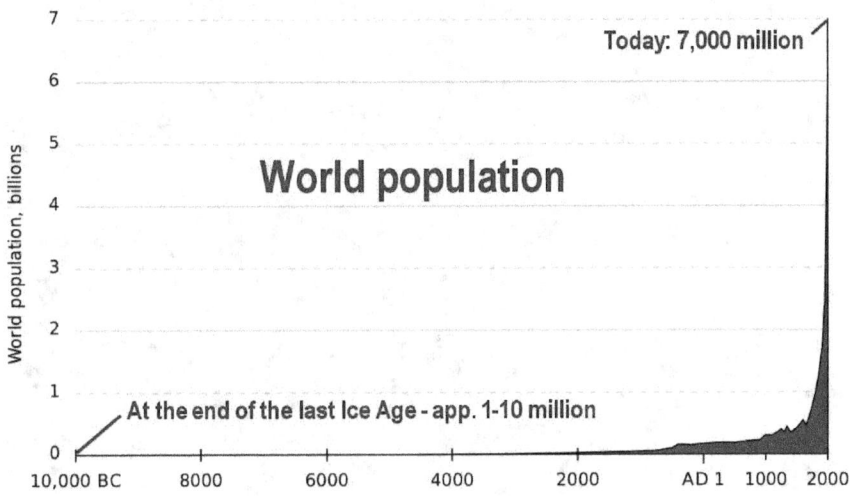

The small size of the world population at the end of the last glacial period, which is estimated to have been in the range of 1 to 10 million, indicates that living is possible in the tropics only under this type of inactive Sun, but that it won't be a picnic by any means. Food resources would have been sparse during the last Ice Age, perhaps consisting mostly of fish.

It is possible for seven billion people to live

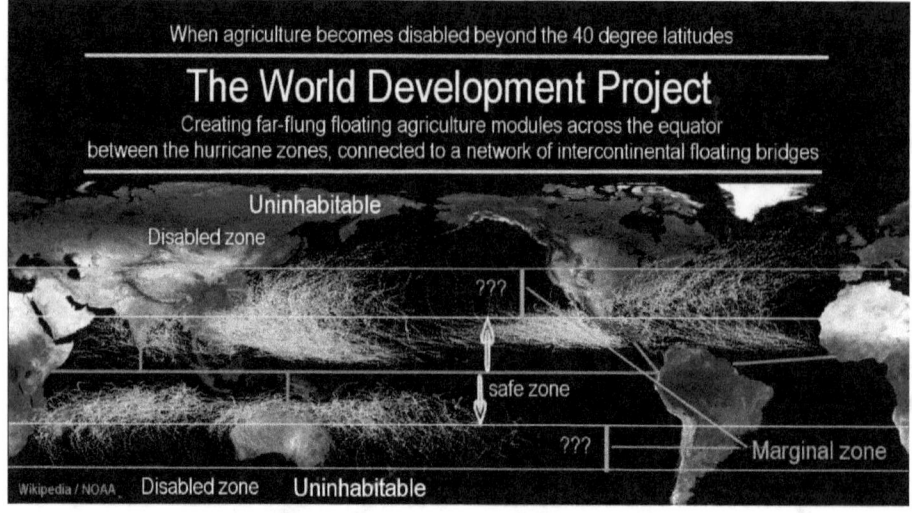

It is possible for seven billion people to live under such harsh conditions, and live richly. But this is not possible on the natural platform that can support but a few. It is only possible on an artificially created, energy intensive, high technology platforms operating in the tropics, interwoven to some degree with large extent indoor agriculture in artificial environments and lighting. These infrastructures can be easily created in automated, high-temperature, industrial processes, together with thousands of new cities and new industries, most of them becoming located afloat across the equatorial seas. In the course of the implementation, a completely new high-temperature industrial revolution would unfold, with basalt as the feed stock. It would revolutionise economics, and with free high quality housing and public facilities, it would revolutionize culture in every respect. We could create the brightest renaissance under an inactive Sun. Anything less would be insufficient. But will we raise ourselves up to that level and become human beings in the highest sense that our humanity enables? Will we grasp the only option we have and live? I think we will, and we will do it before our Sun joins the rank of the 300 billion inactive stars in the galaxy.

That the red-dwarf stars are inactive stars

Fomalhaut, 25 ly distant, 15x brighter than Sun
Credit NASA, ESA, and the Digitized Sky Survey 2*
Acknowledgment: Davide De Martin (ESA/Hubble)

NASA/Hubble, Red dwarf, Proxima Centauri 4.2 ly distant

That the red-dwarf stars are inactive stars becomes evident by comparison. Active stars are typically surrounded by a large blue atmosphere, perhaps of highly activated hydrogen, whereas the inactive stars tend to be surrounded by a darker, more reddish glow, perhaps of less activated hydrogen. The active, brilliant stars are thereby, as by their nature, of an entirely different class.

Stars are classed by luminance into 6 categories

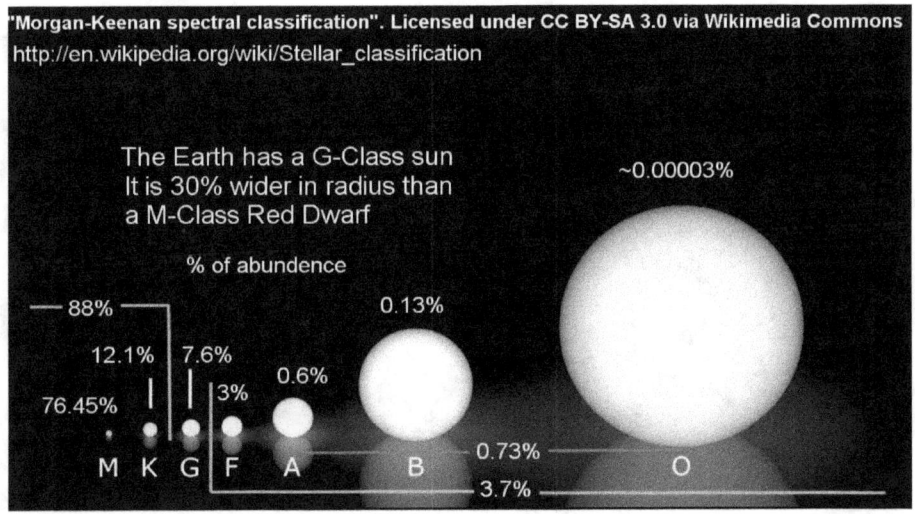

The Earth has a G-Class sun
It is 30% wider in radius than
a M-Class Red Dwarf

% of abundence

~0.00003%

— 88% —
12.1% 7.6%
76.45% 3%
0.6%
0.13%
0.73%
3.7%

M K G F A B O

The stars are classed by luminance into 6 categories, that can be
grouped to two groups. One group contains the really active stars,
type A, B, and O, with a surface temperature ranging upwards to
60,000 degrees. The star, Fomalhaut, from the previous image, is
located on the low end of this group.

The group of extremely active stars is small in numbers. About 3
billion stars belong to this group. These brilliant stars are most likely
the ones to be observed by the automated star mapping systems
that observe 100s of thousands of stars simultaneously.

The other group combines the lesser brilliant stars. This group
contains the Sun of the Earth. And stars with a slightly lower to
slightly higher surface temperature than our Sun. There are about
90 billion stars in this group.

These lesser, still active stars, like our Sun

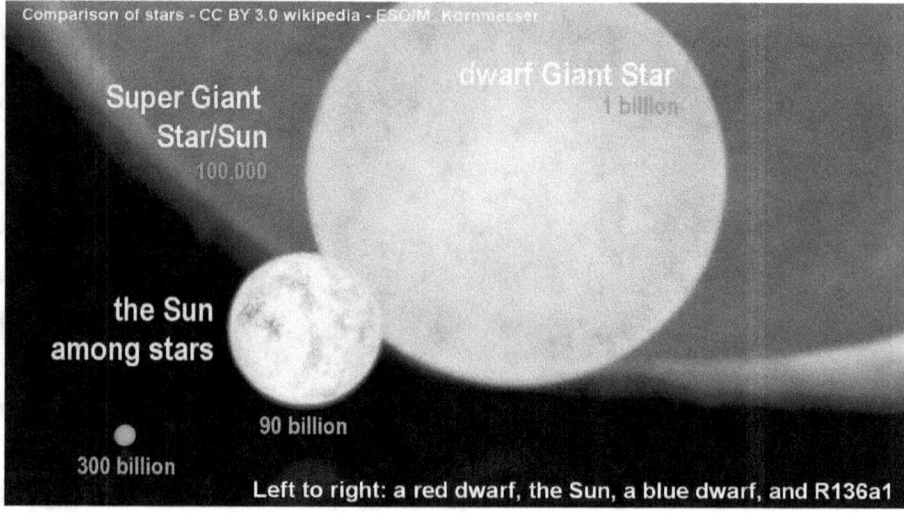

Comparison of stars - CC BY 3.0 wikipedia - ESO/M. Kornmesser

Super Giant Star/Sun
100,000

dwarf Giant Star
1 billion

the Sun among stars

90 billion

300 billion

Left to right: a red dwarf, the Sun, a blue dwarf, and R136a1

These lesser, still active stars, like our Sun, tend to be more vulnerable to going inactive and becoming red stars, being less bright. They are not of the type that is routinely observed, by reasons of their large number and their comparative low luminosity, so that no one would likely notice individual stars of this group going inactive, and if so, the observation would then most likely be recognized as noise in the system, since inactive stars are deemed impossible under the hydrogen-sun theory. It is highly unlikely, therefore, that when such stars go inactive, when they revert to a lower surface temperature, that the event will be noticed or be reported.

Mainstream cosmology has no basis for recognizing inactive stars

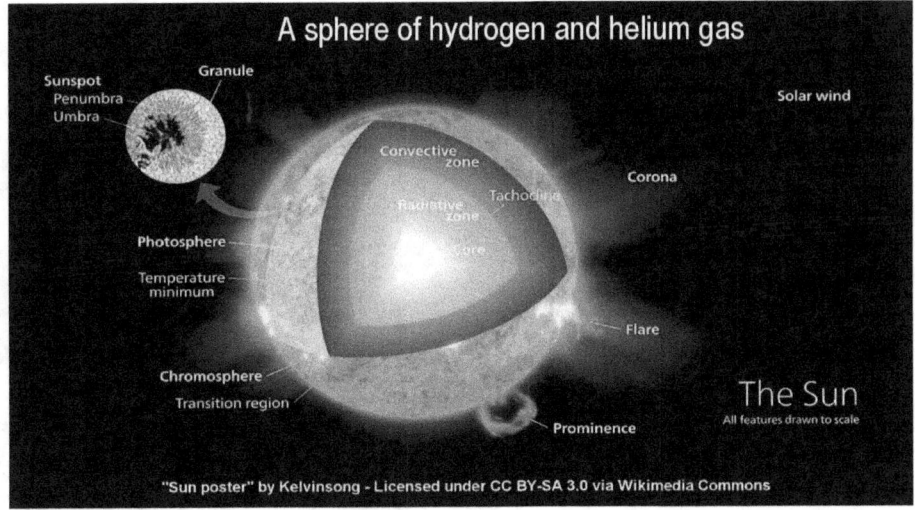

A sphere of hydrogen and helium gas

Sunspot
Penumbra
Umbra
Granule
Solar wind
Convective zone
Corona
Tachocline
Radiative zone
Core
Photosphere
Temperature minimum
Flare
Chromosphere
Transition region
Prominence
The Sun
All features drawn to scale

"Sun poster" by Kelvinsong - Licensed under CC BY-SA 3.0 via Wikimedia Commons

The mainstream cosmology has no basis for recognizing inactive stars. In mainstream cosmology every star, small or large, is deemed to be a hydrogen sun that cannot go inactive, for reasons of it being lit up with nuclear fusion occurring within it. This theory still rules the world, regardless of the fact that a hydrogen sun cannot produce the seamless band of color that we see in the sunlight.

Our Sun would be a thousand times heavier

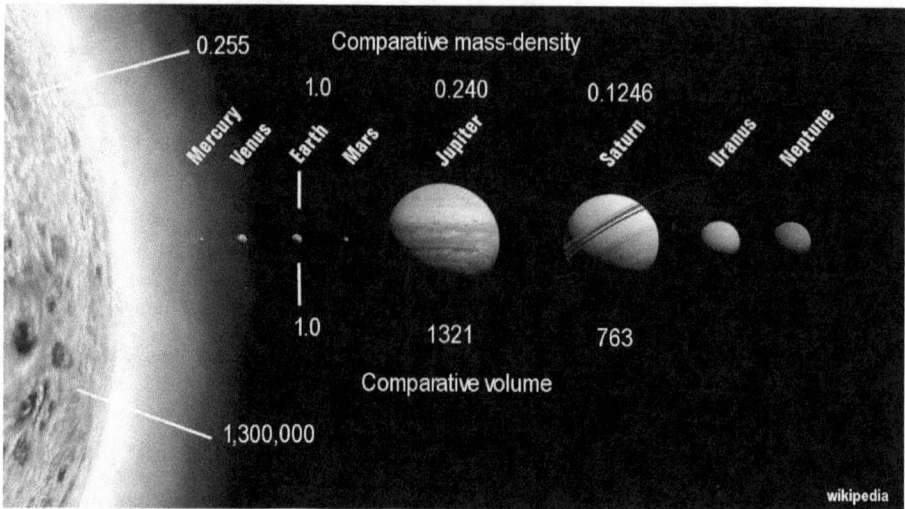

The theory also still rules contrary to the obvious fact that our Sun as a sphere of hydrogen gas, would be a thousand times heavier than it actually is, in comparison with Jupiter and Saturn, nor would a sphere of hydrogen gas of the size of the Sun be able to exist, as the resulting intense gravitational force would crush all atomic structures in its core. In order to rescue the hydrogen-sun theory, the concept of electron degeneracy has been invented that supposedly counteracts gravitational pressure acting on atomic structures. The theory supposedly enables the compression of the hydrogen gas at the core of the Sun to a 200-times greater density than water, without the atoms becoming crushed.

The star UY Scuti, that is 1700-times larger

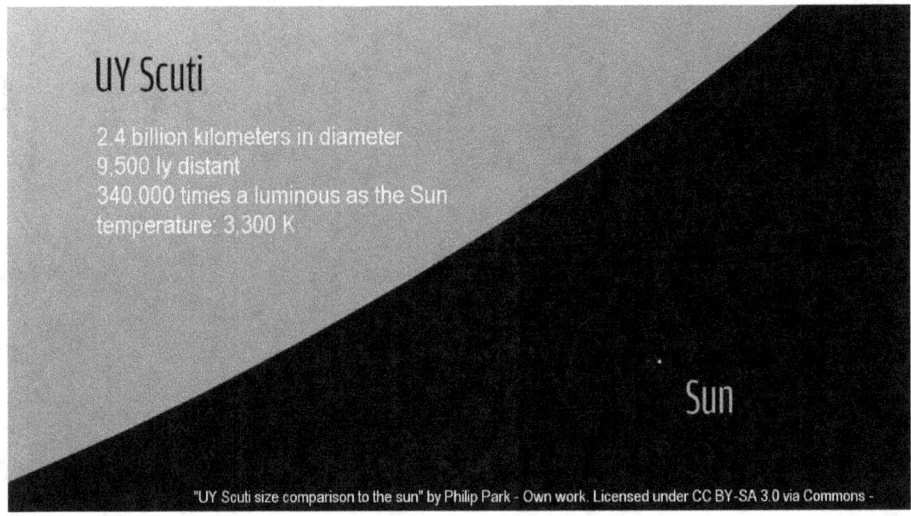

UY Scuti

2.4 billion kilometers in diameter
9,500 ly distant
340,000 times a luminous as the Sun
temperature: 3,300 K

Sun

"UY Scuti size comparison to the sun" by Philip Park - Own work. Licensed under CC BY-SA 3.0 via Commons -

The hydrogen-sun theory becomes even more impossible when one considers the existence of giant stars, like the star UY Scuti, that is 1700-times larger in diameter than our Sun is. A gas sphere of this gigantic size can evidently only be upheld with magic. A sphere of plasma of this immensely huge size, of course, would be able to exist quite naturally. It would be upheld by the repelling electric force of the protons in plasma. The electric force is one of the strongest forces in the universe. It is 39 orders of magnitude stronger than the force of gravity. With the plasma-sun concept, for which the stars themselves stand as evidence, the impossible paradoxes that pervade mainstream cosmology fall by the wayside. No magic is required in the real world, or epicycles, or fudge factors.

Mainstream cosmology champions impossible paradoxes

Mainstream cosmology champions a number of impossible paradoxes that are routinely explained away with magic. For example, no physical principle exists outside the plasma cosmology, for the solar planets to orbit their Sun in a tight ecliptic plain as they do. Exotic causes are cited in mainstream cosmology.

Force the stars in a galaxy to align themselves

Likewise, no physical principle exists in mainstream cosmology that would force the stars in a galaxy to align themselves into the thin ecliptic disks formation that we behold. In plasma cosmology the phenomenon is recognized as a basic electromagnetic phenomenon that has been replicated in laboratory experiments. No magic is active here.

All the stars are deemed to be orbiting the galactic center

Stellar orbits arround the galactic center are NOT possible under Keper's laws of gravitational mechanics. However, no other perception is allowed. Thus new mythical 'epicycles' are imagined to make the modern doctrine plausible.

In addition, in mainstream cosmology all the stars of a galaxy are deemed to be orbiting the galactic center. This is an impossible concept that Kepler's laws of orbital mechanics render totally impossible. But why would anyone care about that? In mainstream cosmology the impossible happens by magic wherever required to keep the doctrines satisfied. In this case the impossible is deemed to be possible by the application of the magic of super-massive black holes, dark matter, and dark energy, which no one has ever seen or can see, and are not even theoretically possible.

The famous Hertzsprung-Russel Star-Data diagram

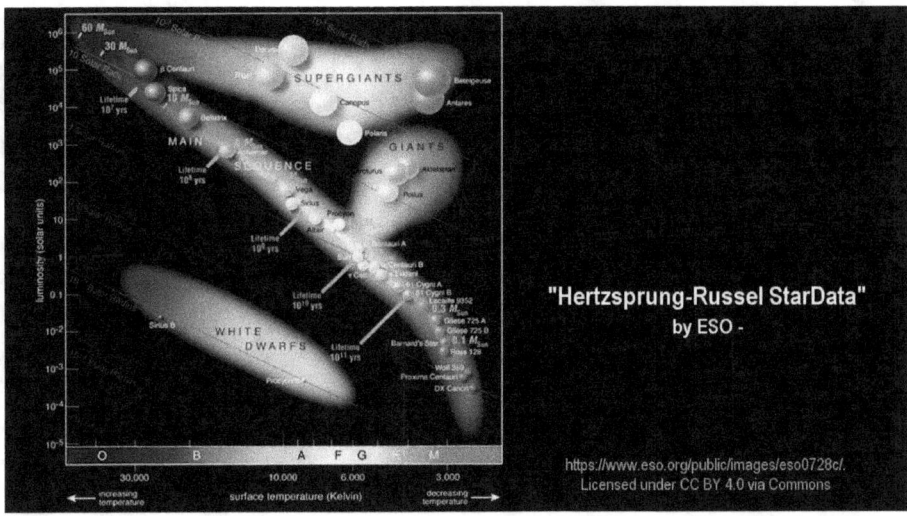

This means that what we see expressed in the stars reflects a totally different operational platform in Plasma Cosmology. In Plasma Cosmology stars can go inactive, and apparently have done so, even among the big stars. Going inactive, of course, doesn't mean a complete turn-off, but means instead that their potential high-intensity state no longer happens. The famous Hertzsprung-Russel Star-Data diagram is arranged by the star's surface temperature, from right to left. The grouping of star types that is shown here, arranged from M on the right for low temperature stars that include the red dwarfs, all the way to O on the left, for the super-intense high-temperature stars that have a surface temperature of up to 60,000 degrees. At first glance it appears that a linear progression exists, whereby the larger stars are inherently hotter stars. The progression holds true up to a point.

Of the active stars, the star Sirius

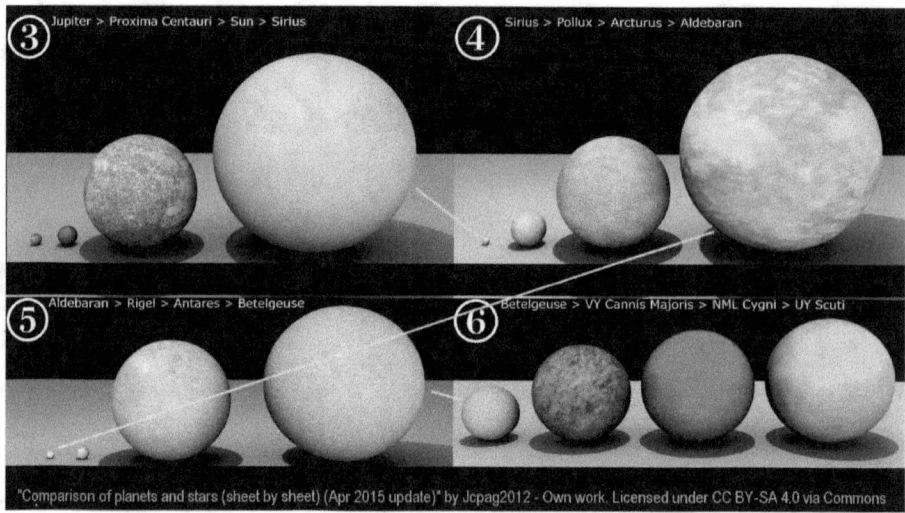

In the sequence of comparison of progressively larger stars, shown here, only two intensely active stars appear after the Sun, with the remaining types standing as examples of what one might term, inactive stars. Of the active stars, the star Sirius in panel 3, stands as an example for a range of active stars that are larger than the Sun and have a higher surface temperature. The Sun has a surface temperature of 5,800 degrees Kelvin. The larger stars exceed this.

The star Fomalhaut

The star Fomalhaut

Distance: 25 ly
1.8 times the diameter of the Sun
15 times as luminous as the Sun
temperature 8,500 K

Credit: NASA, ESA, and the Digitized Sky Survey 2. Acknowledgment: Davide De Martin

The star that is closest in size to the Sun, is the star Fomalhaut. It is 1.8 times larger in diameter, and has the correspondingly higher surface temperature of 8,500 degrees.

Sirius is twice as large as the Sun

Sirius A and B (artwork)
NASA, ESA and G. Bacon (STScI)
distance 8.6 ly

Sirius A
diameter 2x that of the Sun
25x the luminance of the Sun
temperature 9,900 K

The star Sirius, in turn, is twice as large as the Sun, and has a surface temperature of 9,900 degrees, almost twice that of the Sun.

The star Spica-A

Size of active stars and their surface temperature

Star name	Times the Sun (diameter)	Surface temperature
Sun		5,800 K
Fomalhaut	1.8	8,500 K
Sirius	2.0	9,900 K
Spica B	3.64	18,500 K
Antares B	5.2	18,500 K
Spica A	7.4	22,400 K

For the star Spica-A, which is the larger star of a dual star system and is 7.4 times as big as the Sun, the measured surface temperature is a whopping 22,400 degrees.

A principle appears to be reflected in this progression between the size of stars and their correspondingly higher surface temperature. The progression in temperature seems to indicate that larger stars are able to attract larger volumes of interstellar plasma, which electromagnetic fields focus around them, that the stars consume in surface plasma fusion. The focused plasma becomes more concentrated around the larger stars, by the larger star having a greater electric and gravitational potential, which all play a role in the plasma-fusion dynamics.

The progression breaks down

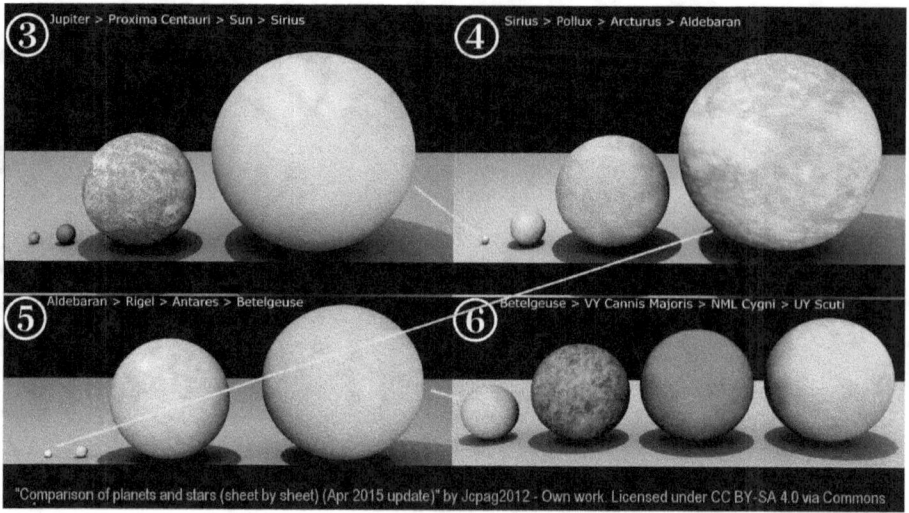

③ Jupiter > Proxima Centauri > Sun > Sirius

④ Sirius > Pollux > Arcturus > Aldebaran

⑤ Aldebaran > Rigel > Antares > Betelgeuse

⑥ Betelgeuse > VY Cannis Majoris > NML Cygni > UY Scuti

The progression breaks down when one considers the still larger stars, Arcturus, in panel 4, and Aldebaran. Both stars a larger in diameter, than the stars in the group represented by Sirius

The star, Arcturus is 25 times larger

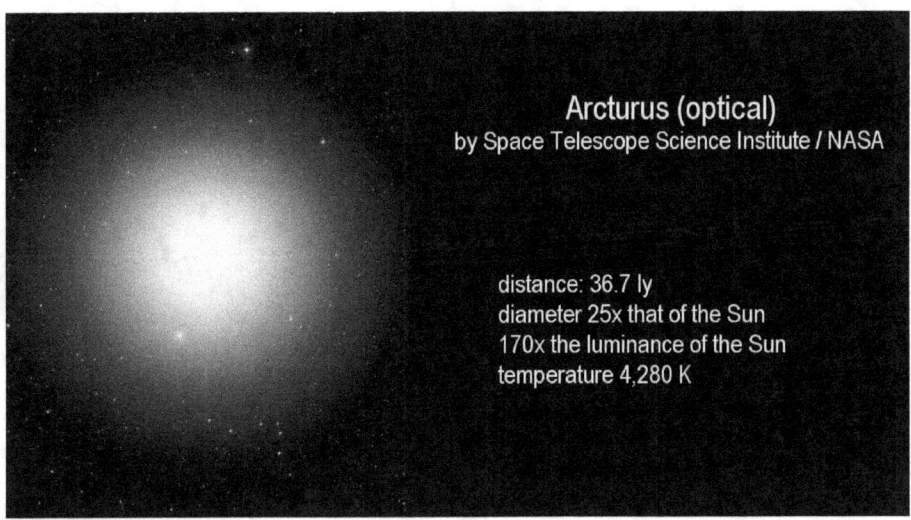

Arcturus (optical)
by Space Telescope Science Institute / NASA

distance: 36.7 ly
diameter 25x that of the Sun
170x the luminance of the Sun
temperature 4,280 K

The star, Arcturus is 25 times larger in diameter than the Sun, but is a cold star with a surface temperature of only 4,200 degrees. The same is evident for the star Aldebaran.

Alderbaran is 44 times larger in size than the Sun

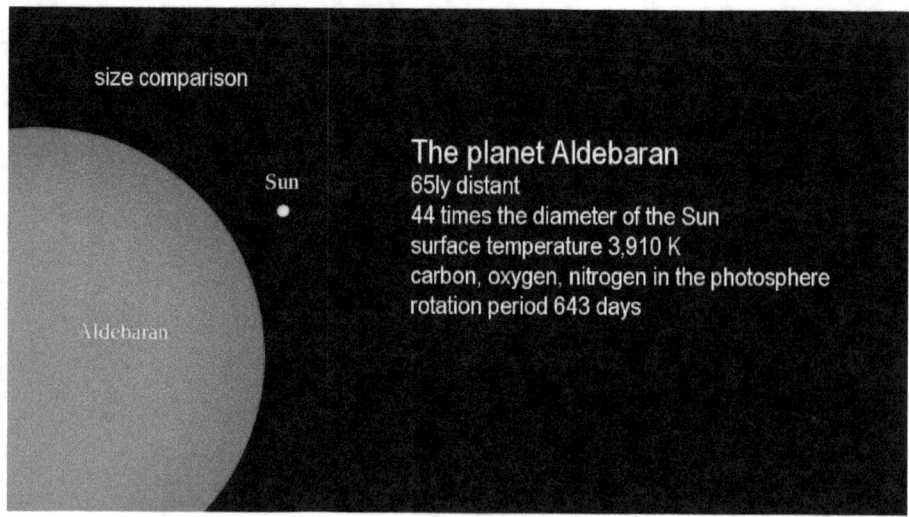

Alderbaran is 44 times larger in size than the Sun, but achieves only a surface temperature of 3,910 degrees.

The progression in surface temperature by size is broken

Size of active stars and their surface temperature

Star name	Times the Sun (diameter)	Surface temperature	
Sun		5,800 K	
Fomalhaut	1.8	8,500 K	
Sirius	2.0	9,900 K	
Spica B	3.64	18,500 K	
Antares B	5.2	18,500 K	
Spica A	7.4	22,400 K	
Arcturus	44	4,286 K	The universal default range
Aldebaran	65	3,910 K	for 'small' inactive stars

The progression in surface temperature by size is broken by these two examples, as if the part of the dynamics that causes intense solar activity no longer functions. Instead of Arcturus, by it being larger in size, having a still higher surface temperature than the 22,400 degrees of Spica-A, Arcturus is a much colder star and Alderbaran even more so. The break in the progression suggests that a systemic failure has occurred in these cases that enables only a lesser type of plasma fusion to occur.

For the Sun, the high-density plasma compression

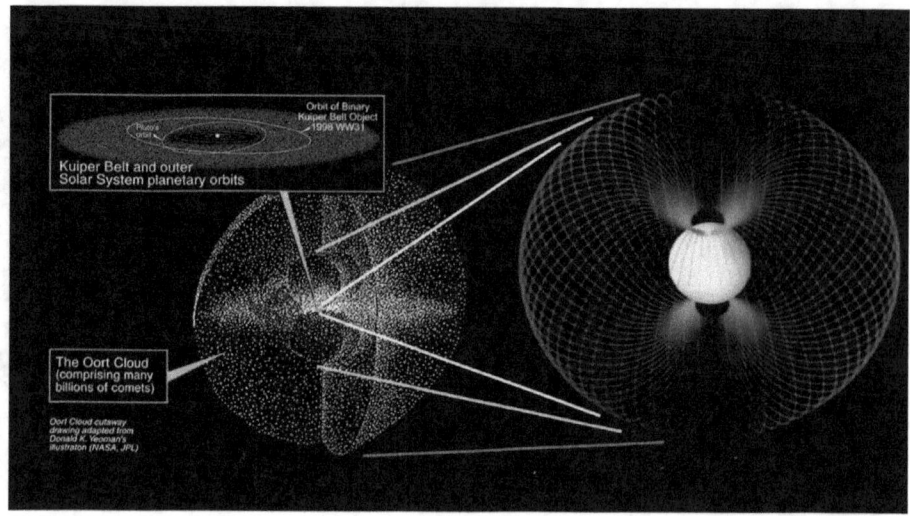

For the Sun, the high-density plasma compression, which enables its 5,800 degrees high temperature plasma fusion to happen, appears to be the product of a 3-fold nested system of primer fields. If one of the three stages of the nested primer fields would fail, the result would be a much-reduced plasma pressure around the Sun, and a much reduced plasma fusion activity occurring on its surface, with a much lower surface temperature being the result of it. The concept of a 3-fold nested system of primer fields is difficult to illustrate and to visualize, but evidence for it does exist.

For the stars Arcturus and Alderbaran

Arcturus (optical)
by Space Telescope Science Institute / NASA

distance: 36.7 ly
diameter 25x that of the Sun
170x the luminance of the Sun
temperature 4,280 K

For the stars Arcturus and Alderbaran one of the stages may have broken down as the result of the diminishing plasma density in the galaxy, or may not have existed.

It evidently takes a lot of plasma

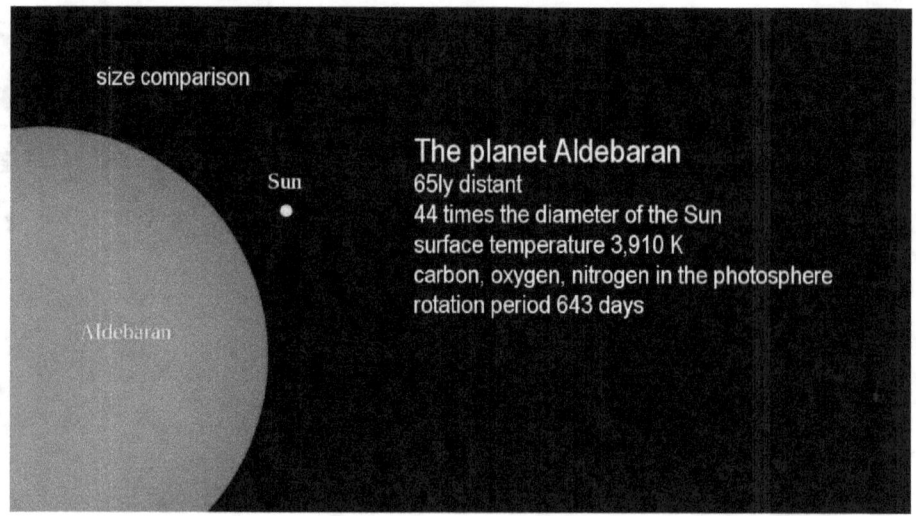

size comparison

Sun

Aldebaran

The planet Aldebaran
65ly distant
44 times the diameter of the Sun
surface temperature 3,910 K
carbon, oxygen, nitrogen in the photosphere
rotation period 643 days

It evidently takes a lot of plasma to power a large star like Alderbaran to its full capacity.

Plasma density may vary regionally across the galaxy

Also the plasma density may vary regionally across the galaxy. We have evidence for that.

The low surface temperature of Arcturus and Alderbaran

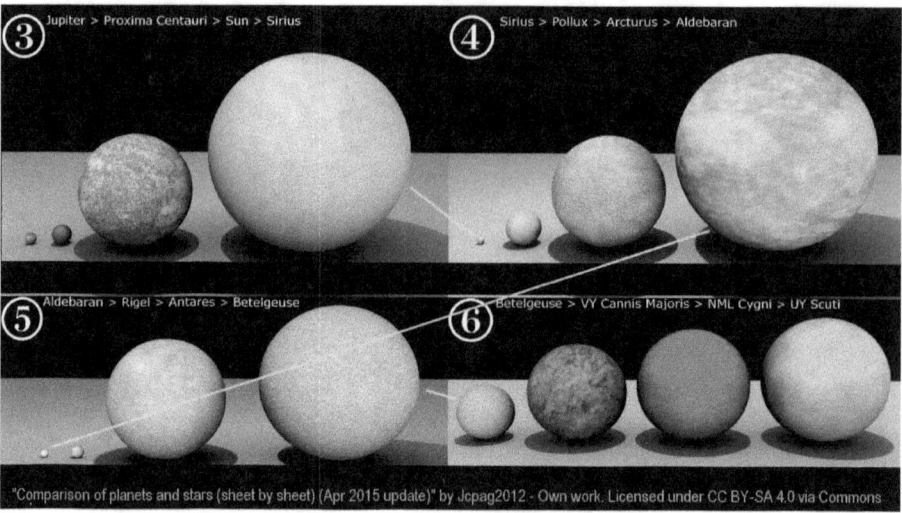

The low surface temperature of Arcturus and Alderbaran is not necessarily typical for large stars. The exception is the star, Rigel, that we find in panel 5. This star is 3 times larger than Arcturus and almost twice as large as Alderbaran. It should be a dim star by its size in comparison with Arcturus and Alderbaran. But it isn't. It is intensively active, though not quite as active as it should be for an active star of its size.

Rigel is nearly 80-times larger than the Sun

The giant star Rigel is nearly 80-times larger than the Sun and has an estimated surface temperature of 12,000 degrees, located at a distance of 880 light years from us. Rigel proves that giant stars can be highly active stars if the plasma density exists. For Rigel, there might not be enough of it. A fully active star of its size would have a surface temperature of 50,000 to 60,000 degrees.

Blue stragglers

Quite a few giant, bright blue stars do exist in the galaxy.
Sometimes they are referred to as blue stragglers.

The giant active star, Rigel, is of the B class

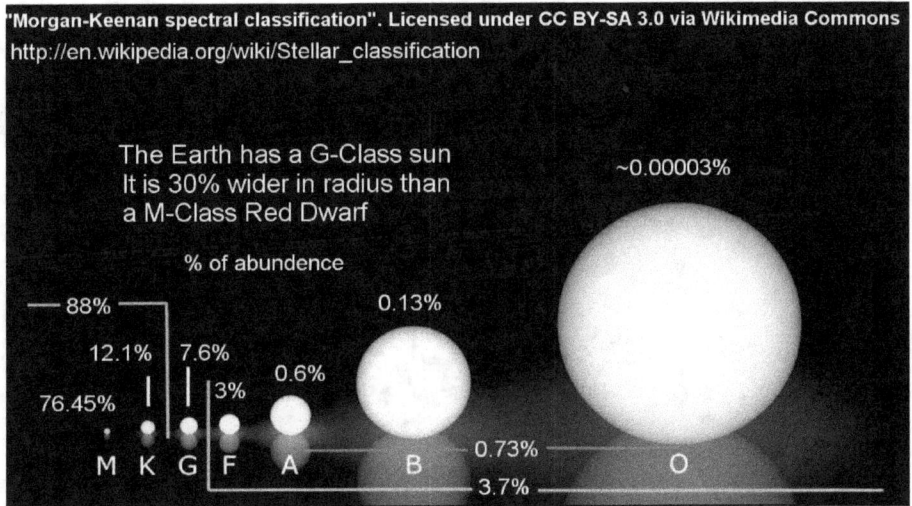

The Earth has a G-Class sun
It is 30% wider in radius than
a M-Class Red Dwarf

~0.00003%

% of abundence

— 88% —

0.13%

12.1% 7.6%

0.6%

76.45%   3%

M K G F A   B

0.73%

3.7%

O

In the classification by surface temperature, the giant active star,
Rigel, is of the B class of roughly half a billion stars. The O class of
stars with surface temperature exceeding 30,000 degrees, typically
into the range of 65,000 degrees, is extremely small. Only 20,000
stars of this group are believed to exist in our galaxy. They are
typically found at the center of a nebula. Their luminosity is several
million times that of the Sun.

Two examples of Class-O stars

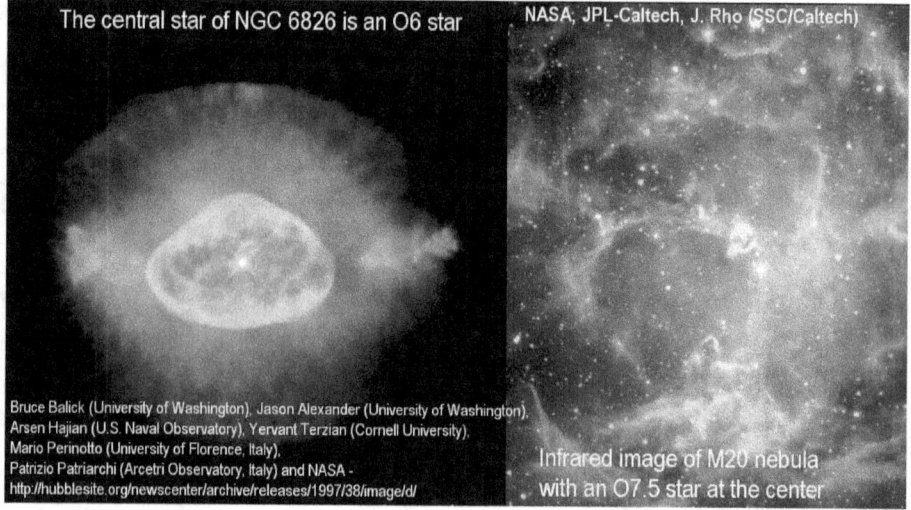

The central star of NGC 6826 is an O6 star

NASA; JPL-Caltech, J. Rho (SSC/Caltech)

Bruce Balick (University of Washington), Jason Alexander (University of Washington),
Arsen Hajian (U.S. Naval Observatory), Yervant Terzian (Cornell University),
Mario Perinotto (University of Florence, Italy),
Patrizio Patriarchi (Arcetri Observatory, Italy) and NASA -
http://hubblesite.org/newscenter/archive/releases/1997/38/image/d/

Infrared image of M20 nebula
with an O7.5 star at the center

In the two examples of Class-O stars, the atomic material that make up the nebula are evidently the plasma-fusion products of the super-active super-giant stars. The atomic elements in the nebula emit light by their interaction with the plasma flowing in the sphere of the nebula, typically in the form of solar wind and interstellar plasma flowing through the nebula.

The O-class star that has created the nebula

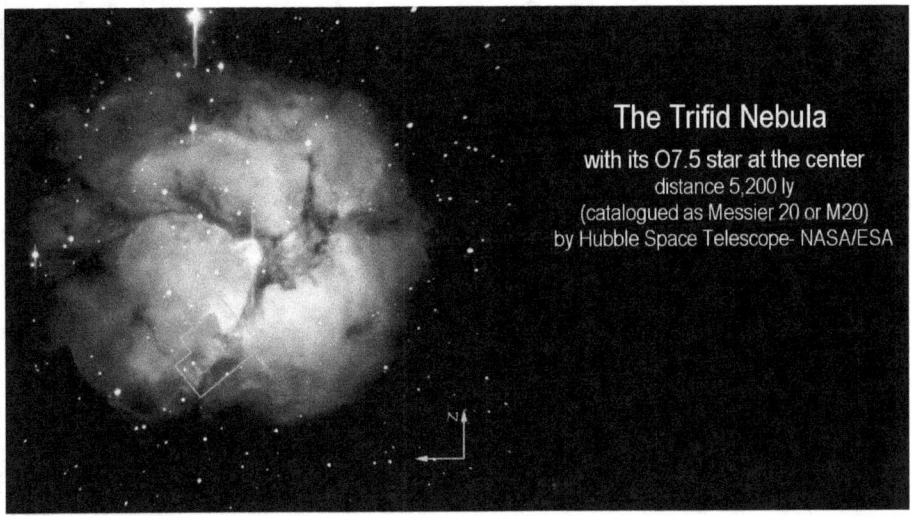

The Trifid Nebula
with its O7.5 star at the center
distance 5,200 ly
(catalogued as Messier 20 or M20)
by Hubble Space Telescope- NASA/ESA

The O-class star that has created the nebula shown here, shines bright and clear across 5,200 light years of space. But not all giant stars are of this category. Many of the giant stars are inactive stars.

Inactive stars in the range of 3,000 to 4,000 degrees

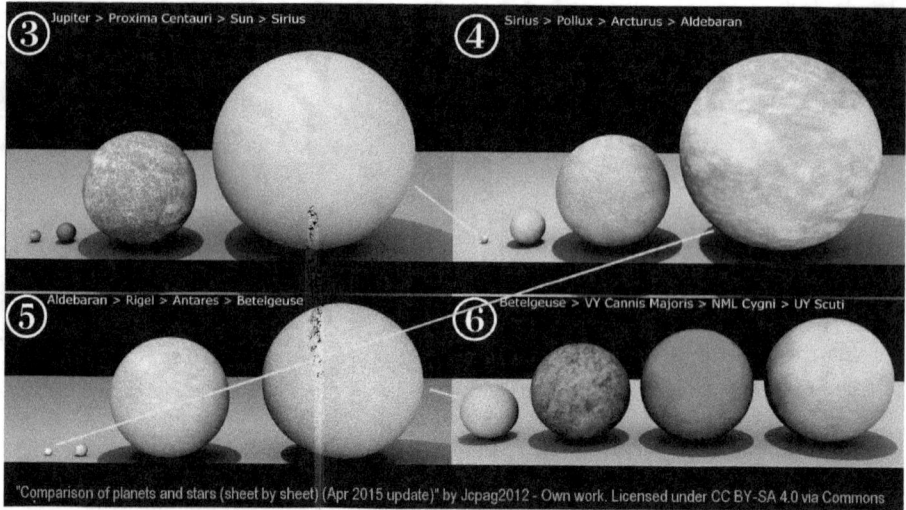

All the giant stars that are shown here from Antares on, in panel 5, to the super-giant UY Scuti, in panel 6, are inactive stars. While these stars span a large range in size, they share one common feature. Their surface temperature is 'cold,' in the range of 3,000 to 4,000 degrees, which is also the typical surface temperature of the red-dwarf stars.

The main star of the Antares system, Antares A

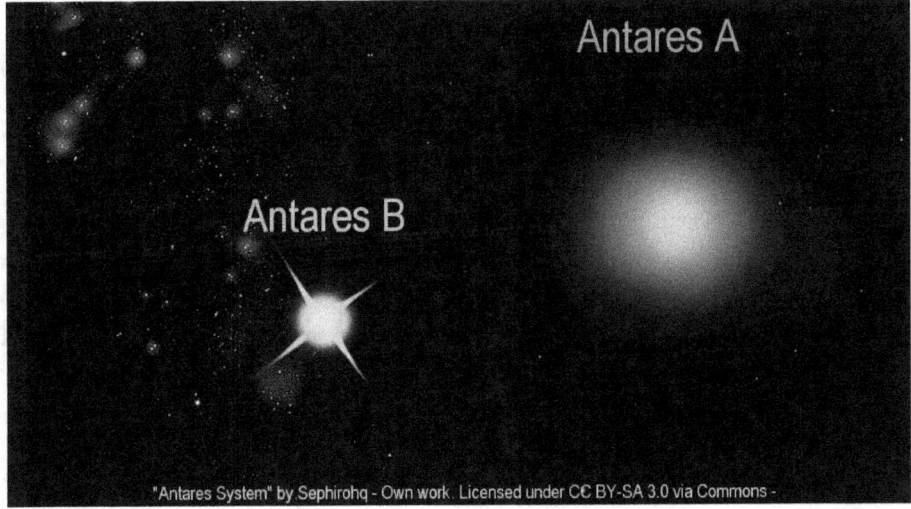

"Antares System" by Sephirohq - Own work. Licensed under CC BY-SA 3.0 via Commons -

For example, the main star of the Antares system, Antares A, is a supergiant star that is 880 times larger than the Sun, but has a surface temperature of only 3,400 degrees, which is typical for inactive stars. It is so dim that it is barely visible across its 550 light years' distance. Antares B, in comparison, which is an active companion star, is only 5.2 times as large as the Sun but has a surface temperature of 18,500 degrees that is typical for what may be termed, active stars. In the image shown here, the size of the smaller Antares-B is dramatically exaggerated, perhaps to illustrate the visual difference between an active and inactive star. In its active days, Antares-A might have been a superstar.

Antares-A is 300 million kilometers in diameter

Sun R = 0.7 million km

Orbit of Mars
R ≈ 227 million km

Orange Star
Arcturus R = 20 million km

Red Giant
Antares R ≈ 300 million km

Antares Size comparison

550 ly distant
temperature 3,400 K
10,000 times as luminus as the Sun
diameter 883x that of the Sun

Antares-A is 300 million kilometers in diameter. Its huge size dwarves the Sun, but, because of its low surface temperature of 3,400 degrees, as an inactive star, the giant star is only 10,000 times as luminous as the Sun, with a dim red hue, instead of the millions of times greater luminosity that it might have once had as an active star.

The star Betelgeuse is of the same category

The star Betelgeuse
via ESO's Very Large Telescope

Distance: 430 ly
1180 times the radius of the Sub
120,000 as luminous as the Sun
temperature 3,300 K

The star Betelgeuse is of the same category, together with
countless more like it. Betelgeuse is an inactive giant that is 1180
times larger in diameter, than the Sun, and is thereby only 120,000
times more luminous with its low surface temperature of only 3,300
degrees, instead of the million times greater luminosity it could
have.
In mainstream cosmology, the inactive super-giants are deemed to
be stars that have consumed their hydrogen fuel, and have begun
burning helium instead. The concept renders them as candidates
for future supernova explosions when their helium fuel is
exhausted, whereby the stars are deemed to contract. In plasma
cosmology, however, where stars are recognized as spheres of
plasma, which do not burn themselves out and collapse, the
supernova phenomenon is far-less exotic. It may be caused by a
wayward planet colliding with the plasma-sphere of a sun. In the
collision, its atomic elements would become crushed by the star's
internal gravity in a chain-reaction nuclear-fission event that would

be similar to a planet size atomic bomb going off.

I am bringing this up, because we might be seeing the same type of nuclear fission process happening on the very small scale, in the giant inactive stars. An inactive star attracts atomic elements with its gravity, from its surroundings, and crushes their atomic structures with gravitational pressure within its plasma sphere. While this is possible, it would likely result in a lower temperature than 3,300 degrees, which means that the giant stars are powered by low-level plasma fusion instead.

An example for a star burning by nuclear fission

Antares might be an example for a star burning by nuclear fission of atomic matter drawn from its surrounding. It might also be an example of low-level plasma fusion happening in a low-density plasma environment in which the synthesized atomic elements hang around the star like a cloud.

When our own Sun goes inactive

Ice Age of the dimming Sun in 30 years

www.ice-age-ahead-iaa.ca

In this sense, when our own Sun goes inactive, with which the next Ice Age begins, it would remain powered by interstellar plasma, but in a radically less-compressed form around it.

The Earth will not loose its Sun

Our Sun
as an inactive star

Our Sun may appear diffused, which is typical for inactive stars, and a yellowish tint, which is typical for 4,000 K surface temperature

The Earth will not loose its Sun then, when the Sun goes inactive, but have a cooler, dimmer, and more diffused Sun with a surface temperature of roughly 4000 degrees Kelvin.

The hyper-giant star, UY Scuti

Much the same can be said about the hyper-giant star, UY Scuti, that stands out brightly in the star fields across a distance of 9,500 light years. This star makes the term, gigantic, seem small. It has a diameter of 2.4 billion kilometers, but its surface is cold. Its surface temperature is a mere 3,300 degrees that is typical for a star being lit up by low-intensity plasma fusion, or by nuclear fission.

It is the star's enormous size that makes it 340,000 times as luminous as our Sun, even while it is technically an inactive star. As an active star, it would have millions of times the luminance of the Sun. Its surface temperature would then be acceding 50,000 degrees.

The star Rigel may be at the border line

Size of active stars and their surface temperature

Star name	Times the Sun (diameter)	Surface temperature	
Sun		5,800 K	
Fomalhaut	1.8	8,500 K	
Sirius	2.0	9,900 K	
Spica B	3.64	18,500 K	
Antares B	5.2	18,500 K	
Spica A	7.4	22,400 K	
Rigel	79	12,000 K	
Antares	883	3,400 K	The universal default for large inactive stars
Betelgeuse	1180	3 300 K	
UY Scuti	1700	3 300 K	

In this context, the star Rigel may be at the border line. The plasma density around it is probably too low to enable its full potential, but after that, each of the large stars that stand as examples here, are completely inactive, all being powered to nearly the same temperature regardless of their size, either by low-level plasma fusion, or nuclear fission, or both.

When our Sun goes inactive

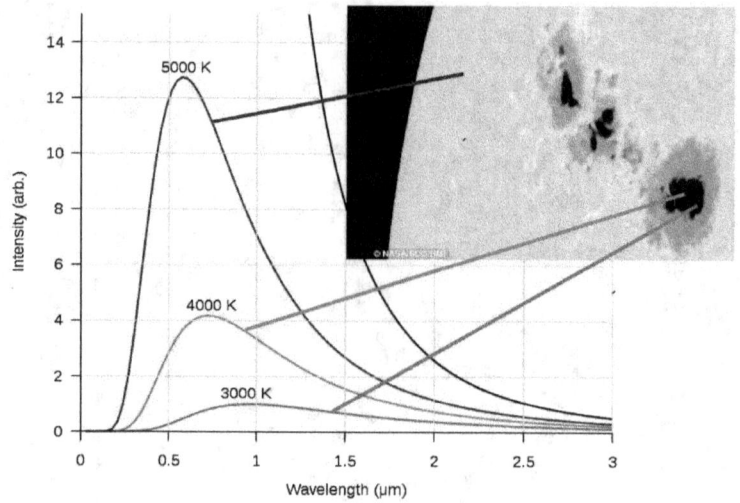

When our Sun goes inactive, with its surface temperature becoming reduced from the present 5,800 degrees to the 4000 degrees' range, a 70% reduction in radiated energy would result that would be similar to what we see in the umbra of the sunspots. The reduction would make the tropics slightly colder than the regions at the 70-degree latitude presently are. Some form of outdoor agriculture would therefore likely be able to continue there, maybe with enhanced lighting. In practice a large portion of the agriculture would have to be carried out in indoor facilities with artificial environments, artificial sunlight and artificially enhanced CO_2 density, and so on. This is in essence what the stars are teaching us.

Plasma-density in our galaxy has diminished

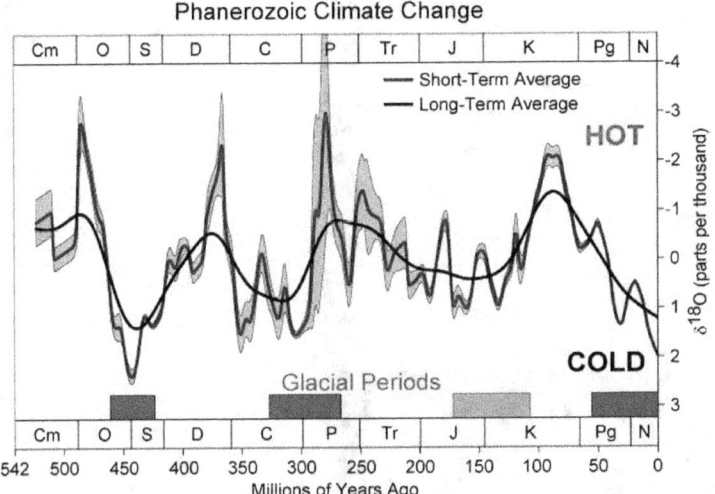

Phanerozoic Climate Change

What we see in the stars illustrates to some degree how deeply the plasma-density in our galaxy has diminished.

We presently experience the consequence of the long-ongoing down-ramping of the plasma density in our galaxy that we have evidence of. The down-ramping began roughly 100 million years ago. The galaxy that we see today with its vast array of inactive stars, is the result of the massive down-ramping in the galaxy. The measured long-term climate variations on Earth appears to be representative of the cyclical plasma density variations in the galaxy. And those variations are big.

Two-thirds along Antarctica froze up

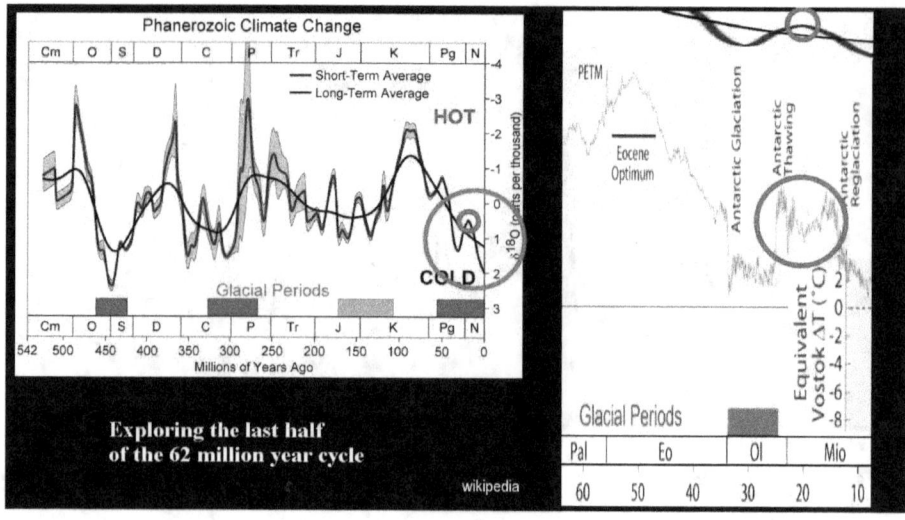

Exploring the last half of the 62 million year cycle

Two-thirds along the down-ramping to the present state, Antarctica froze up, then thawed out again when the shorter cycle peaked, and a few million years later it froze up once more and has remained frozen. These are huge climate effects on the Earth. They are evidently the result of huge causes that affect the entire galaxy, and the stars within the galaxy. Stars going inactive appear to be the natural consequence of the long-term down-ramping in the galaxy. Our Sun is caught up in the dynamics of the presently diminishing galactic system. The resulting effects are obviously large, which we cannot escape from, but which we can adjust our living to.

We will see the photosphere simply go dimmer

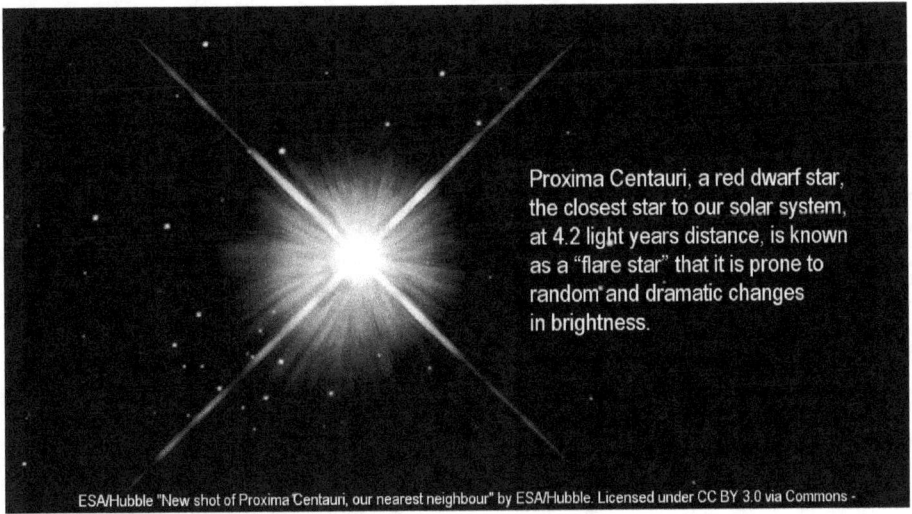

Proxima Centauri, a red dwarf star, the closest star to our solar system, at 4.2 light years distance, is known as a "flare star" that it is prone to random and dramatic changes in brightness.

When our Sun goes inactive, its operating photosphere will likely become transformed to a lower intensity state, rather than vanish as the innermost primer fields collapse that focus high-density plasma onto the Sun.

Most likely we will see the photosphere simply go dimmer, and the space around the Sun become fuzzier, as the solar winds won't sweep the synthesized atomic elements away as efficiently as they do now. Some form of solar wind will likely continue after the photosphere transforms itself.

Should the photosphere vanish completely, the previously synthesized atomic elements would no longer flow away with the solar wind, but would fall back onto the Sun by gravitational attraction. When the attracted elements fall deep into a Sun, a point will be reached when the increasing gravitational pressure will crush the atomic structures of the attracted elements. In the resulting nuclear fission process the previously invested binding energy would become released. The released energy would create a luminous fuzzy shell within the original plasma sphere. Both potentials are possible, though at the present stage the nuclear-fission potential is extremely unlikely,

Photograph of a very-small red-dwarf star

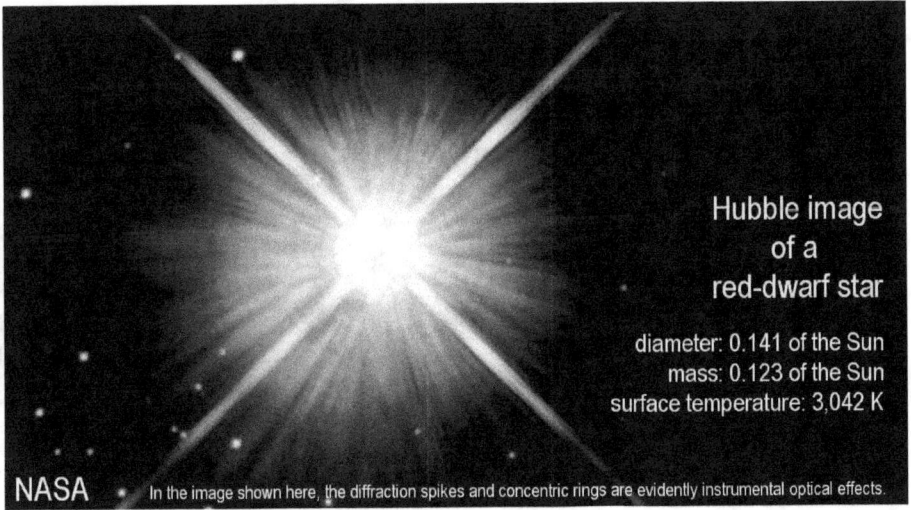

Hubble image
of a
red-dwarf star

diameter: 0.141 of the Sun
mass: 0.123 of the Sun
surface temperature: 3,042 K

NASA • In the image shown here, the diffraction spikes and concentric rings are evidently instrumental optical effects.

The Hubble space telescope has provided us a perfect photograph of a very-small red-dwarf star in the closest solar system to our own, the Centauri system, which is a mere 4.2 light years distant. Is the tiny sphere that we see here, of one of the smallest of the red-dwarf stars, a sphere where nuclear fission takes place? We see the star surrounded by atomic material that glows dimly in the star's plasma sphere. Or do we see in this image an example of a tiny star that is powered by low-level plasma fusion? Most likely, that's what we see.

A tangled network of a vast array of plasma streams

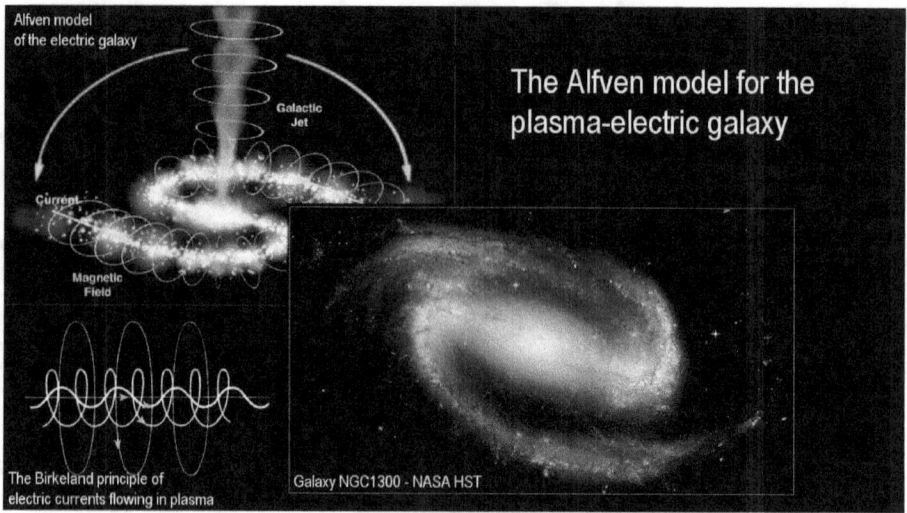

However, the existence of the nuclear-fission powered stage of a star, cannot be ruled out, regardless of its original size. Our galaxy is a tangled network of a vast array of plasma streams that are always in motion and twisting in the spiral arms by the principle of Birkeland currents.

When a star looses its connection

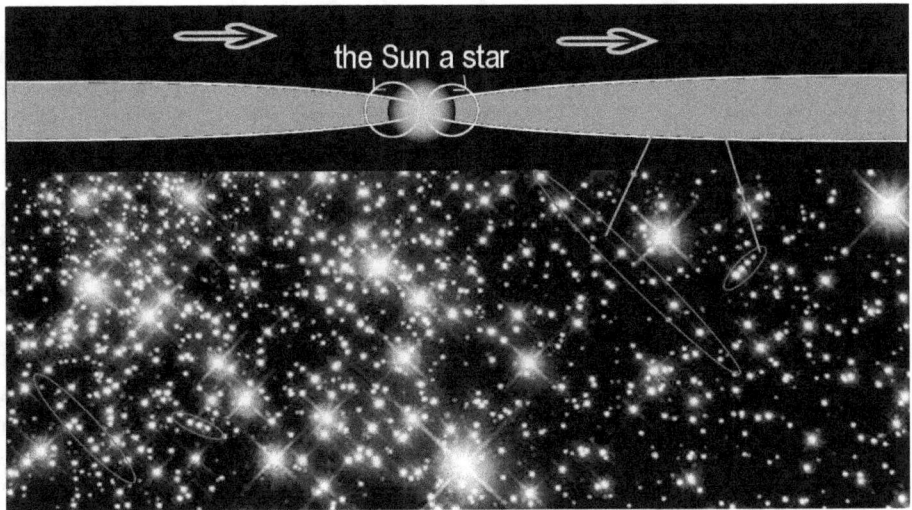
the Sun a star

When a star looses its connection to the interstellar plasma
streams, its plasma fusion stops. Nuclear fission would then be the
only energy source a star would have, until the star would be
reconnected again with the galactic network of plasma streams.
In its disconnected state, the nuclear-fission process would render
the deactivated star, dark, and smaller in apparent size than its
original size, and entropic in nature. Any star that is not externally
powered, is inherently entropic. The entropic dim star would then
consume the atomic material that it has created earlier during its
active state, and when all that would be consumed, the dim star
would become a white-dwarf star.

The white-dwarf star stage

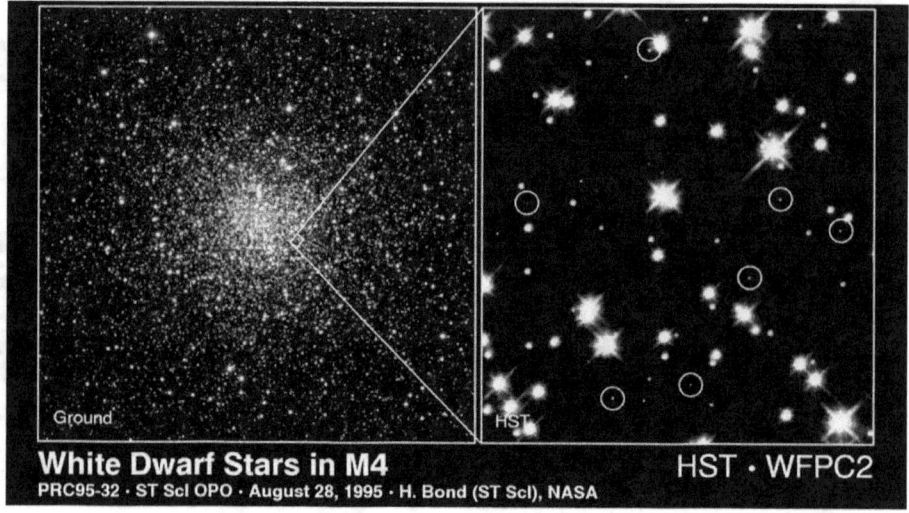

White Dwarf Stars in M4
PRC95-32 · ST ScI OPO · August 28, 1995 · H. Bond (ST ScI), NASA

HST · WFPC2

The white-dwarf star stage might have been reached in some places. Some areas in our galaxy have quite a few of them. Our Sun might have been a disconnected white dwarf once around 700 million years ago, when, as it has been theorized, the Earth froze up completely, and had remained a snowball for a few tens of millions of years. That's just a theory.

Examples of white-dwarf stars are encircled in this image. At the star's dead stage, the pinprick of light appears to be caused by a type of synchrotron radiation that emits light under conditions of extreme plasma pressure, without any atomic elements being involved for the emission of light.

In mainstream cosmology, the white-dwarf star is regarded to be a burnt-out star past the helium stage that has lost thereby its ability to main its nuclear fusion process within. It is believed that electron degeneracy then allows its atomic material to condense, and with it its remaining thermal energy. It is believed that a white dwarf is as dense as if all the mass of the Sun was packed into the sphere of the Earth, which then would glow brightly until it would cool into oblivion. Of course, the electron-degeneracy theory is just a theory that like an epicycle is needed to uphold the Hydrogen-Sun theory in the first place, which has become a doctrine in modern time that stands like a giant in denial of the plasma cosmology concept that is actually supported by a large body of evidence.

Another form of white-dwarf star is also possible in plasma cosmology, in a different manner, which may be what is actually being observed. The mysterious white-dwarf star might simply be a brown-dwarf star that exists in a region of high-density interstellar plasma, in which it becomes an active star. This possibility is the most likely one that we see evident here.

The principle of orbiting stars

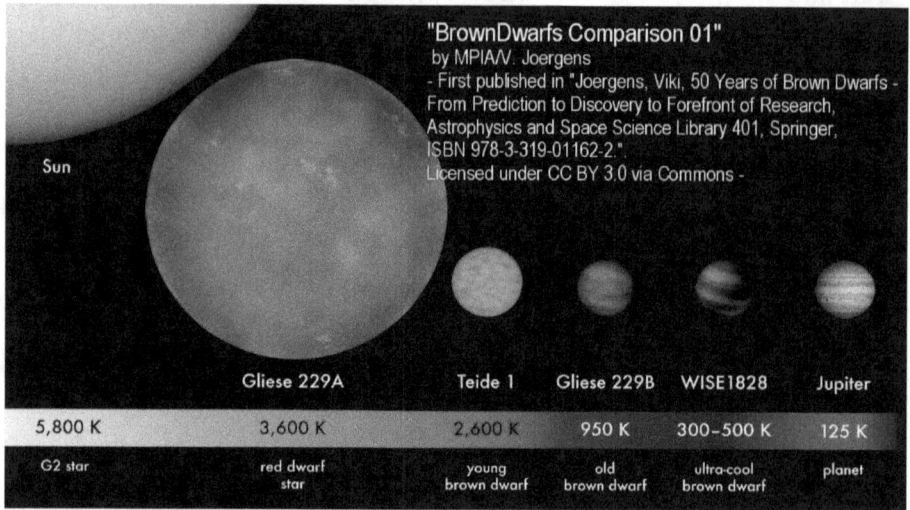

Sun	Gliese 229A	Teide 1	Gliese 229B	WISE1828	Jupiter
5,800 K	3,600 K	2,600 K	950 K	300–500 K	125 K
G2 star	red dwarf star	young brown dwarf	old brown dwarf	ultra-cool brown dwarf	planet

Evidence for such a case exists in the Gliese 229 binary system. In this binary-star system, we have a brown-dwarf star orbiting a red-dwarf star. It illustrates the principle of orbiting stars.

The reason is inherent in plasma physics

The large red dwarf, in this example, is the parent star Gliese 229A. It is 69% as large in diameter than the Sun, and is 59% as large in mass, with a surface temperature of 3,600 degrees. The small companion star, Gliese 229-B, in comparison, is roughly the size of Jupiter. It is a mere 10% of the size of the Sun. Its low-level plasma fusion generates a surface temperature of less than a thousand degrees.

Here a problem would begin if the small brown star did not have the gravitational pressure within it, to cause the fissioning of the atomic elements to happen, that it attracts and would surely attract in close proximity to another star, which in part it would also synthesize itself. As a consequence, the atomic elements would accumulate within such a star, which would render these stars extremely-high-density stars, which they are. The small brown stars are believed to contain 50 times more mass than Jupiter. Such an extremely dense mass in atomic elements would cause the star to eventually fission in a supernova explosion. But this doesn't happen. The reason is inherent in plasma physics.

High-mass brown-dwarf star WISE1828

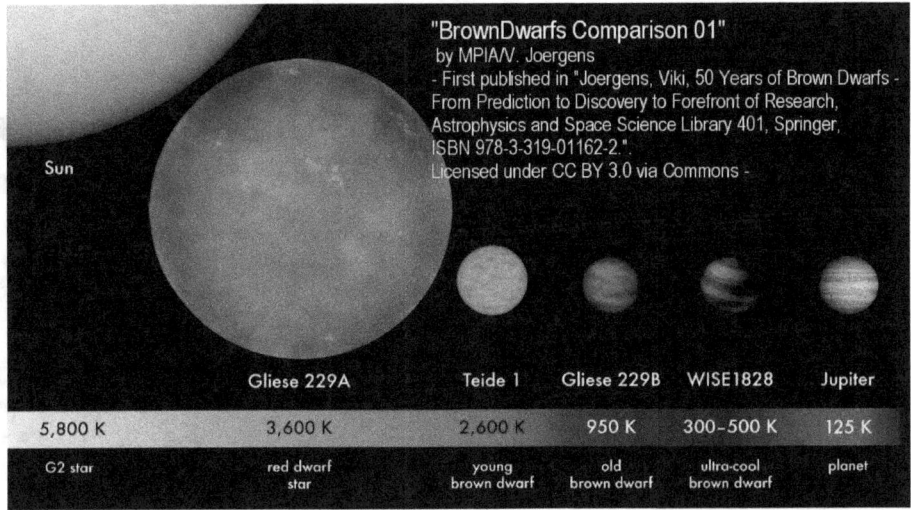

Sun	Gliese 229A	Teide 1	Gliese 229B	WISE1828	Jupiter
5,800 K	3,600 K	2,600 K	950 K	300–500 K	125 K
G2 star	red dwarf star	young brown dwarf	old brown dwarf	ultra-cool brown dwarf	planet

In the case of a high-mass brown-dwarf star, like WISE1828, that may have a 50 times greater mass than Jupiter contained in the same volume as Jupiter, the star would have a mass density that is 12 times greater than that of the Earth, which has the highest mass density of all the planets. The comparison renders the brown-dwarf star many times denser than led, even denser than uranium.

How is this enormous mass-density possible for such a small star? In atomic physics the tight packaging would be a miracle that's not possible by any means. But in plasma physics, this extreme mass density is possible, and is evidently quite natural.

Inherently low mass-density in atomic structures

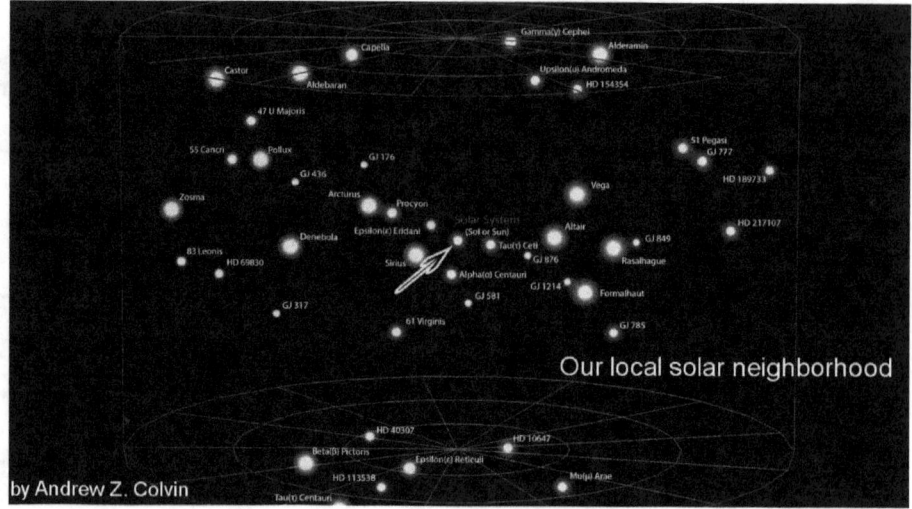

Our local solar neighborhood

by Andrew Z. Colvin

More than 1,800 brown-dwarf stars have been identified in our stellar neighbourhood. They are too small and numerous to be shown here. The reason why it is possible for these high-density plasma stars to exist, can be recognized when one explores the inherently low mass-density in atomic structures.

An atom is a dynamic structure

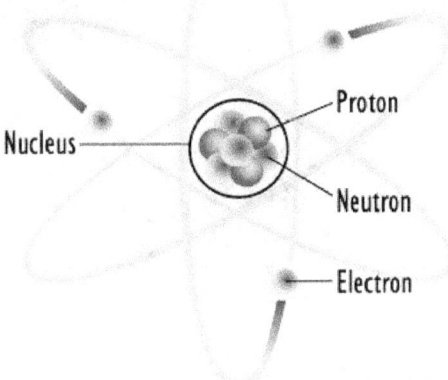

Nucleus

Proton

Neutron

Electron

An atom is a dynamic structure in which the swarming of electrons around a nucleus neutralizes the repelling force of the electric field of the protons at its center. The resulting packaging gives the atom a specific mass density.

An atom is 100,000 times larger than its parts

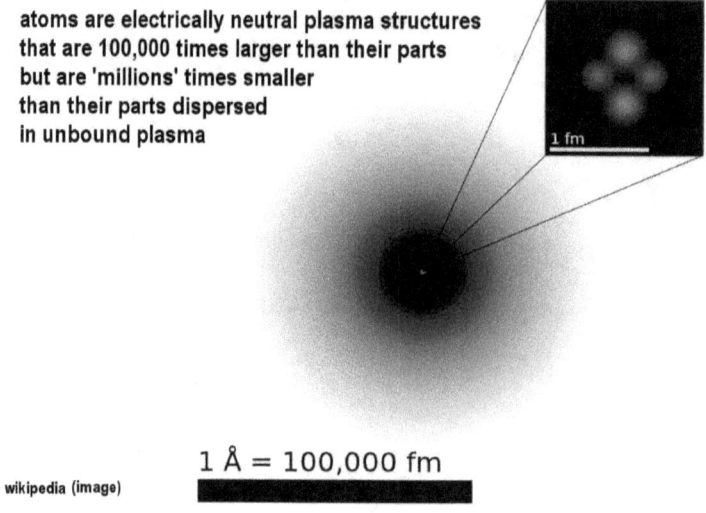

atoms are electrically neutral plasma structures
that are 100,000 times larger than their parts
but are 'millions' times smaller
than their parts dispersed
in unbound plasma

1 fm

$$1 \text{ Å} = 100,000 \text{ fm}$$

That the resulting mass-density in an atom is extremely low,
becomes apparent when one considers that an atom is typically
100,000 times larger than the sum of its parts, which are the
protons and the electrons that form the package.
In plasma, however, which is made up of protons and electrons in
free roaming form, the electrons are not energized enough to
perform specific functions. They simply remain free flowing.

When the electron density in the plasma is high

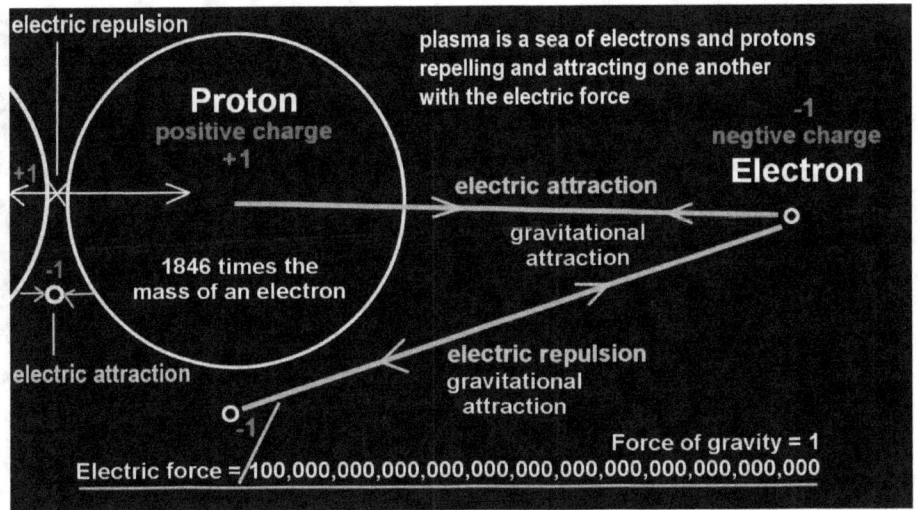

In unbound plasma, of course, the protons all repel one another by the electric force of their equal polarity. The repulsion, however, is counteracted by the electric force of the electrons that have the opposite polarity. When the electron density in the plasma is high, all the plasma particles can exist together much-more tightly packed than in atomic form. This principle enables the tiny brown-dwarf stars to have an enormous mass density.

A tiny star orbiting a large star

The planet Sirius
and its companion
a white-dwarf star

NASA - Hubble Space Telescope

Another, similar case, of a tiny star orbiting a large star, in this case an extremely active star, is the case of the star, Sirius. We see a tiny star orbiting the large star. The tiny star shines brightly in the dense plasma environment that powers the large star. This tiny pin-prick star is regarded to be a high-mass white dwarf. It is extremely unlikely that a white dwarf would be found at this close distance to an active star. It is far-more likely that we see a brown dwarf in this image, that is intensely activated in the high-density plasma sphere that typically surrounds a large active star, as Sirius is. Sirius is twice as large as the Sun and nearly twice as hot.

Any size of star can be formed with plasma concentrations

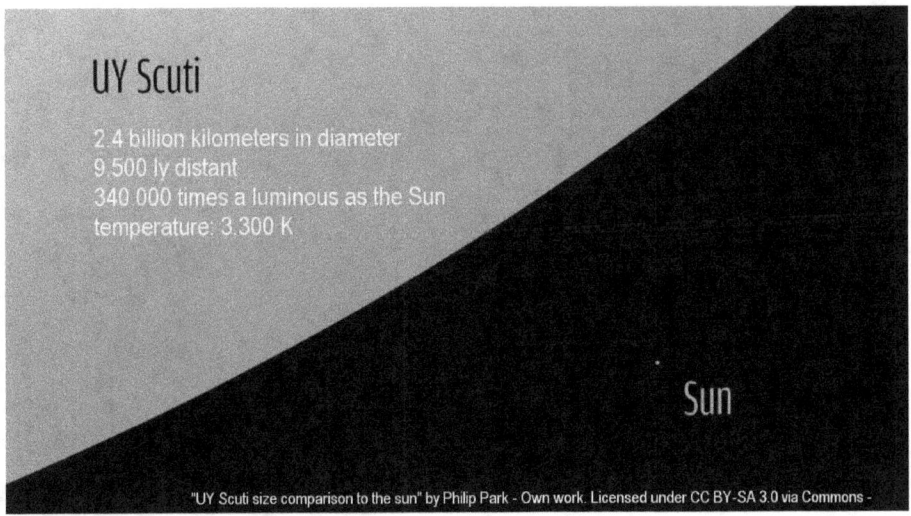

UY Scuti

2.4 billion kilometers in diameter
9,500 ly distant
340,000 times a luminous as the Sun
temperature: 3,300 K

Sun

"UY Scuti size comparison to the sun" by Philip Park - Own work. Licensed under CC BY-SA 3.0 via Commons -

In the plasma universe, any size of star can be formed with plasma concentrations, and have evidently been formed as we see it in the case of UY Scuti. The star is 1700 times as large in diameter than the Sun, which makes it 5 billion times larger in volume, while its mass is believed to be barely 10 times larger than that of the Sun. As one researcher has put it, the mass of this giant star is so thin that it is almost a vacuum.

It is hard to imagine, even in mainstream cosmology, that a star with this extremely low mass density would be able to exist as a hydrogen star powered by nuclear fusion that is caused by gas-compression. Only a miracle could cause that to happen, and to cause the resulting fusion to make the resulting star 340,000 times as luminous as our Sun presently is, with only a 10 times greater mass.

In plasma cosmology no miracle is required

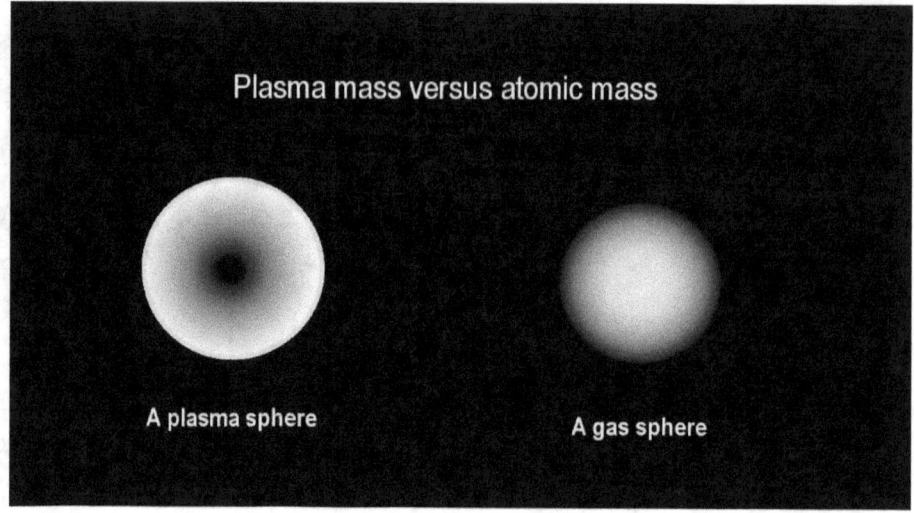

In plasma cosmology, in contrast, no miracle is required for very large stars to exist and to operate.

Under the force of large gravitational pressure at the center of a star, the much lighter electrons in the plasma mix tend to become squeezed out of the interior of the plasma sphere unto its surface. When this happens, the interior expands by the force of proton repulsion that is less counteracted by the diminished electron-density, which asserts an attracting force on the protons. As a consequence, the star becomes larger. By this electric principle, the mass-density of a plasma star increases towards the surface, which is the complete opposite of what we find in atomic mass concentrations that have their greatest density at the center of a sphere.

With a plasma star having its greatest mass-density at its surface, extremely large stars can form that have a relatively small total mass, and operate efficiently.

Surface area functions as a catalyst for interstellar plasma

Even in cases when the large sphere of a star is only able to achieve low-level plasma fusion, that gives it a surface temperature of a mere 3,300 degrees, the resulting large-surface star becomes nevertheless a highly luminous star. It becomes this not by its own power, but because its very large surface area functions as a very large catalyst for interstellar plasma streams. This is how it is possible for a star with 10 times the mass of the Sun, to outshine the Sun 340,000 fold, even while it is almost empty inside.

The principle also applies to our own star, the Sun

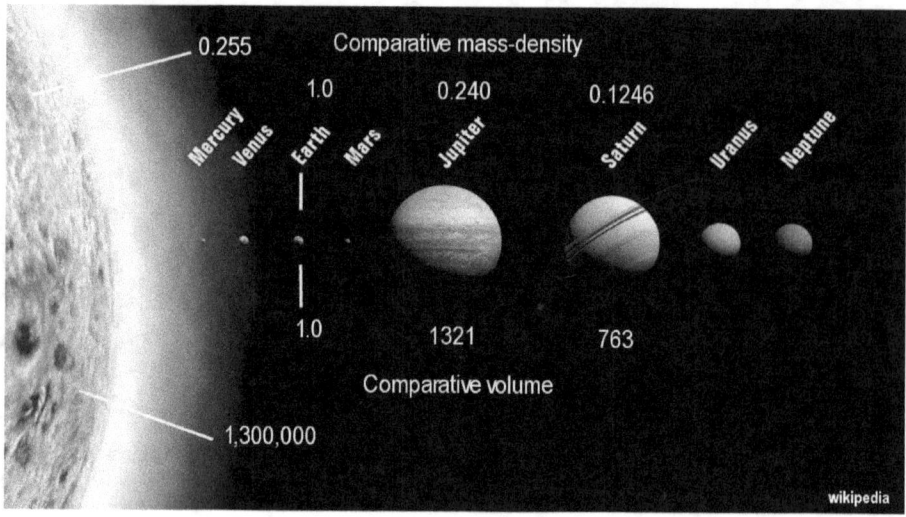

The principle, evidently, also applies to our own star, the Sun. The Sun's mass-density is roughly the same as that of the planet Jupiter. If the Sun was a sphere of hydrogen gas, its mass-density would be a thousand times greater, because of the gravitational compression at the center of the Sun. Jupiter is twice as large in volume than Saturn. Consequently, Jupiter has double the mass-density, because of the greater mass compression, with both being gas planets. By this principle, the Sun should have a thousand times greater mass-density than Jupiter, with it having a thousand times greater volume. But that's not the case. However, with the Sun being a sphere of plasma that is largely empty inside, being essentially but a shell of plasma, the low mass-density that it is known to have, is just about right.

The Sun's plasma shell is dense enough

Size of active stars and their surface temperature

Star name	Times the Sun (diameter)	Surface temperature
Sun		5,800 K
Fomalhaut	1.8	8,500 K
Sirius	2.0	9,900 K
Spica B	3.64	18,500 K
Antares B	5.2	18,500 K
Spica A	7.4	22,400 K

In its presently highly active state, the Sun's plasma shell is dense enough to support surface plasma fusion that heats it up to 5,800 degrees Kelvin. A larger sun, by this principle, would have a denser shell with a greater electron density at the surface, which would enable higher surface temperatures, as is indeed the case.

The resulting low-level default value

Size of active stars and their surface temperature

Star name	Times the Sun (diameter)	Surface temperature	
Sun		5,800 K	
Fomalhaut	1.8	8,500 K	
Sirius	2.0	9,900 K	
Spica B	3.64	18,500 K	
Antares B	5.2	18,500 K	
Spica A	7.4	22,400 K	
Arcturus	44	4,286 K	The universal default range
Aldebaran	65	3,910 K	for 'small' inactive stars

With the Sun being at the low-end in the range of active stars, when the primer fields become disabled that focus interstellar plasma onto it, the Sun will have to work with what the unfocused interstellar plasma streams deliver to it. The resulting low-level default value appears to be in the range of 4000 degrees or less.

Very large stars have lower default values

Size of active stars and their surface temperature

Star name	Times the Sun (diameter)	Surface temperature	
Sun		5,800 K	
Fomalhaut	1.8	8,500 K	
Sirius	2.0	9,900 K	
Spica B	3.64	18,500 K	
Antares B	5.2	18,500 K	
Spica A	7.4	22,400 K	
Rigel	79	12,000 K	
Antares	883	3,400 K	The universal default for large inactive stars
Betelgeuse	1180	3 300 K	
UY Scuti	1700	3 300 K	

Some of the very large stars have lower default values, that are nearly the same across the board. These values might be lower, because there simply may not be enough density in the default plasma background for the giant stars to achieve their full inactive potential. But this shouldn't concern us for the case of the Sun, which is far too small a star for such considerations.

It is enough for us to know that evidence tells

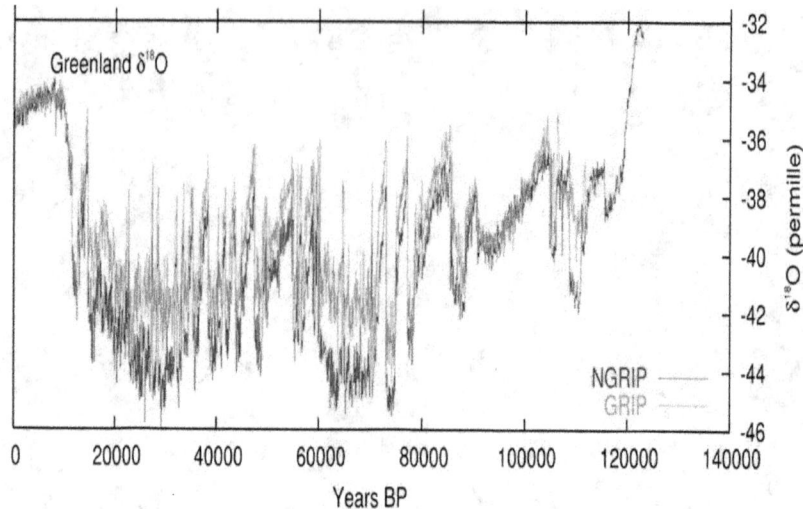

It is enough for us to know that evidence tells us that when our Sun goes into its inactive mode during the glaciation period, its surface temperature will drop to near the 4,000 degrees' level, and that it will become briefly reactivated in intervals of 1470 years, all the way through the 90,000-year glaciation period.

What the Dansgaard Oeschger oscillations indicate

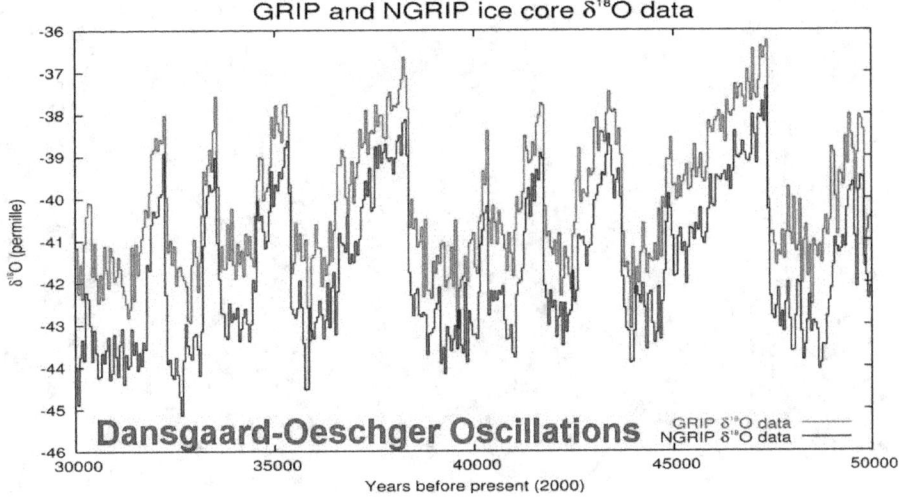

That's what the Dansgaard Oeschger oscillations indicate, has happened in the past. That's what also the stellar dynamics indicate with a high degree to be totally possible, and will likely happen again in the future as it has happened all the way through the previous glaciation cycle.

Evidence supports the inactive stage

Ice Age of the dimming Sun in 30 years

www.ice-age-ahead-iaa.ca

This evidence that one sees presently, rules out the white-dwarf stage and the entropic red-dwarf stage, but supports the type of inactive stage at which the Sun is being powered by interstellar plasma that is less focused on it, because of potentially collapsed primer fields, and is less concentrated around it than it presently is, with a remaining surface temperature of 4,000 degrees or less. That's the bottom line.

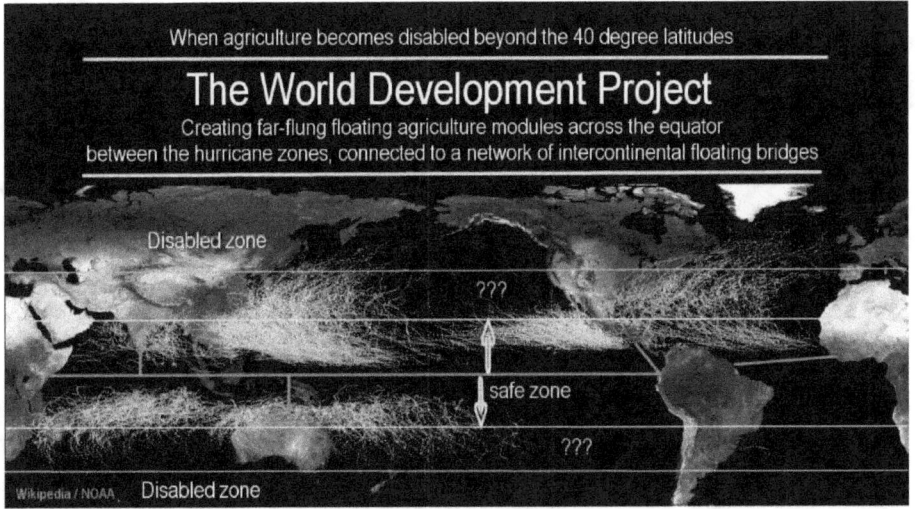

This knowledge gained, is the result of the extraordinary capability that we have as human beings, to explore the principles that operate in the universe, and then to utilize our discoveries as a basis for building our future before it happens.

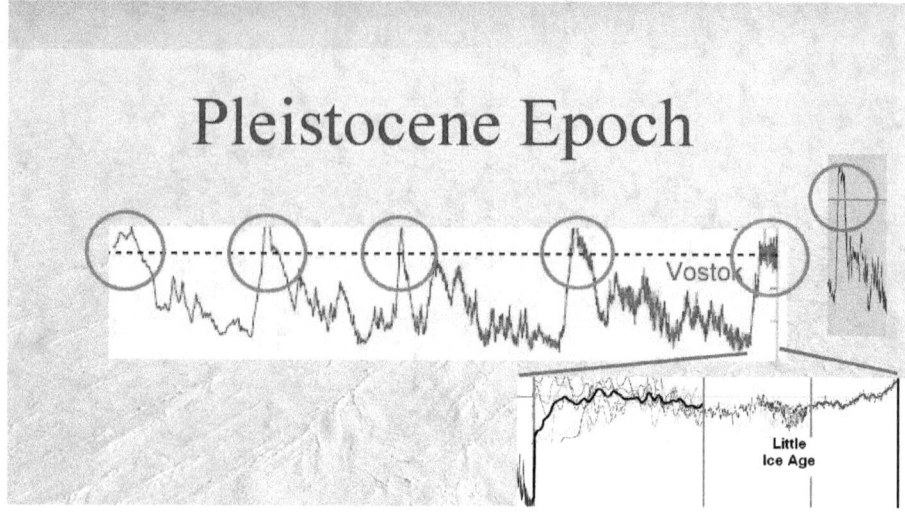

The evidence prompts us to recognize that our star, the Sun, is presently extremely vulnerable to reverting back to its low-intensity inactive state that has been the norm for the last half a million years that we have ice core records for, and that this event of getting back, to the state of normal, will likely occur in roughly 30 years whereby the current interglacial pulse ends that has been our climate home for the last 12,000 years.

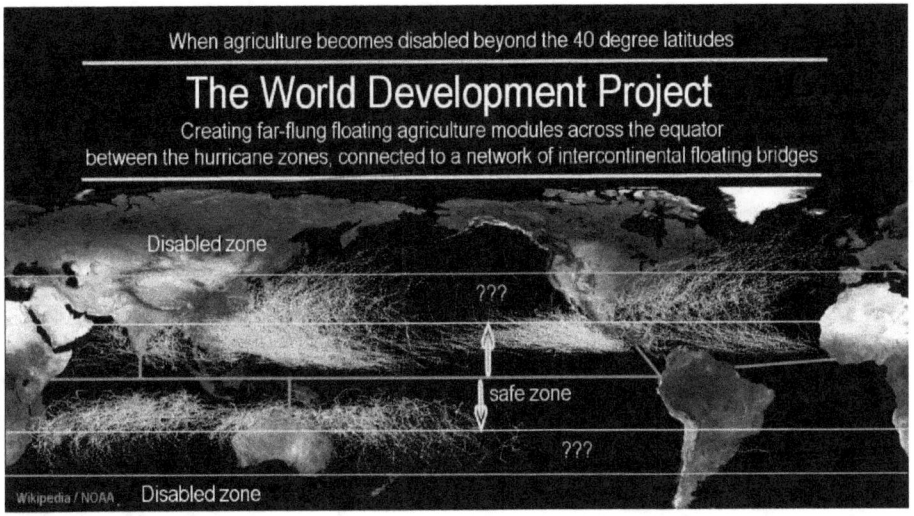

With this consideration we face the point where the academic significance of plasma cosmology gains a life-critical significance in the real world, because, if the academic recognition is not sufficiently achieved to impel the relocation of most nations of the world into the tropics, together with new forms of agriculture and new cities, and this before the phase shift begins, then very few people living today, including their children, will have a chance to remain alive. The entire world becomes radically altered when the Sun becomes a red star and joins the rank of the roughly 300 billion red stars in the galaxy, most of which are red dwarfs that might have been once brightly shining active stars.

The large volume of the presently inactive stars in the 4,000 degrees range in surface temperature, in the field of our galaxy, is not really surprising if one considers that our galaxy is presently at its weakest state in over 400 million years.

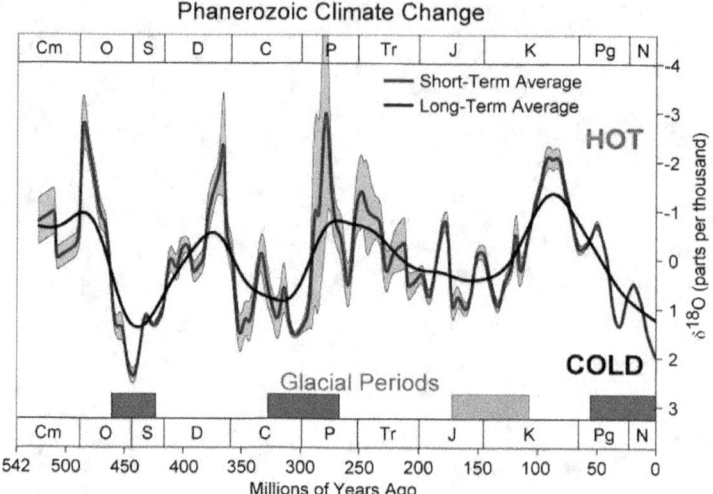

With measurements of oxygen-18 isotope ratios, which are climate sensitive, it has become possible to measure the climate history on Earth, going back in time for almost 500 million years.

The measurements indicate the existence of two very long resonance cycles, that appear to exist in the intergalactic plasma streams that power the galaxy. These resonances together affect the plasma density across the galaxy.

In x-ray and gamma-ray light

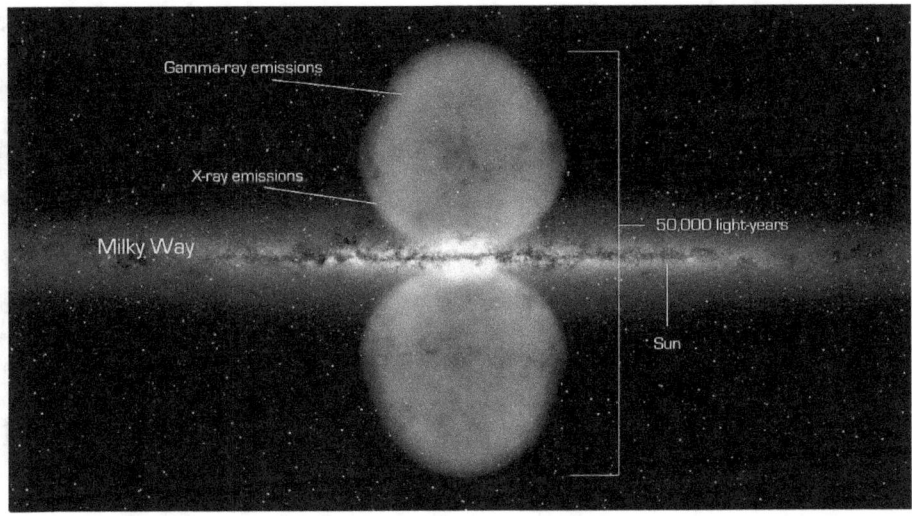

When NASA explored the galaxy in x-ray and gamma-ray light, two gigantic plasma domes became visible.

Galactic-scale confinement domes

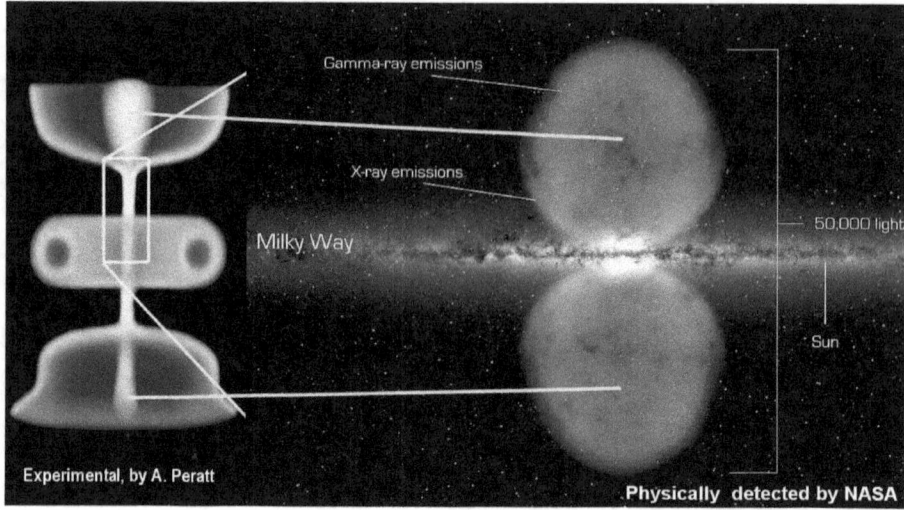

Gamma-ray emissions

X-ray emissions

Milky Way

50,000 light

Sun

Experimental, by A. Peratt

Physically detected by NASA

These domes are amazingly similar to the electromagnetic confinement domes that are visible in plasma discharge experiments, which serve to immensely concentrate plasma streams at their node point. In this sense the plasma domes come to light as galactic-scale confinement domes at the node point of two plasma streams with our galaxy between them.

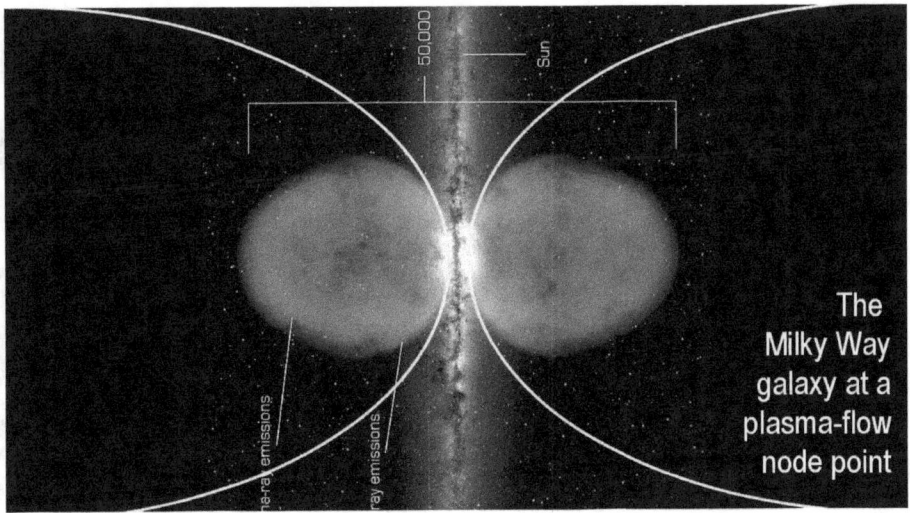

The plasma domes would be formed at the terminal end of two long intergalactic plasma streams, with each having a different resonance.

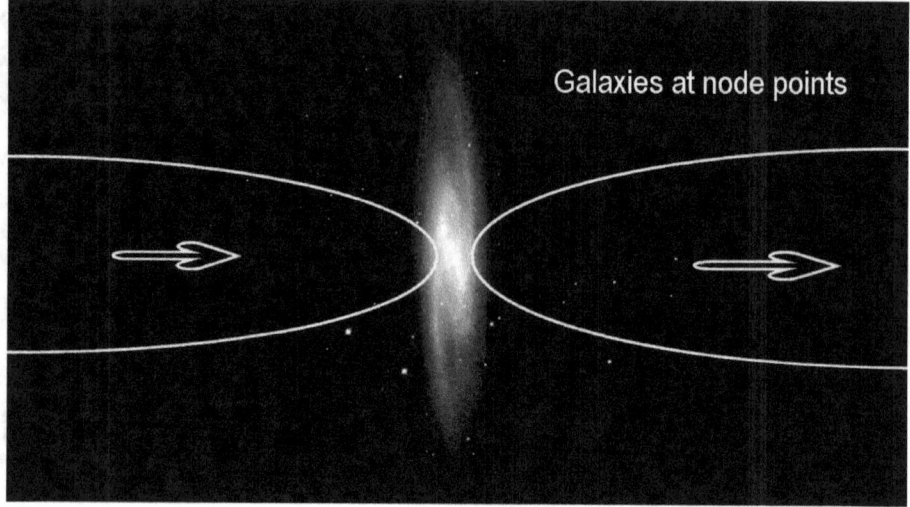

The longest of these plasma streams would have a resonance that is reflected in the 145 million years cycle that has been discovered in the Earth's climate history, and the shorter plasma stream would have a resonance that is reflected as the 31 million years' climate cycle.

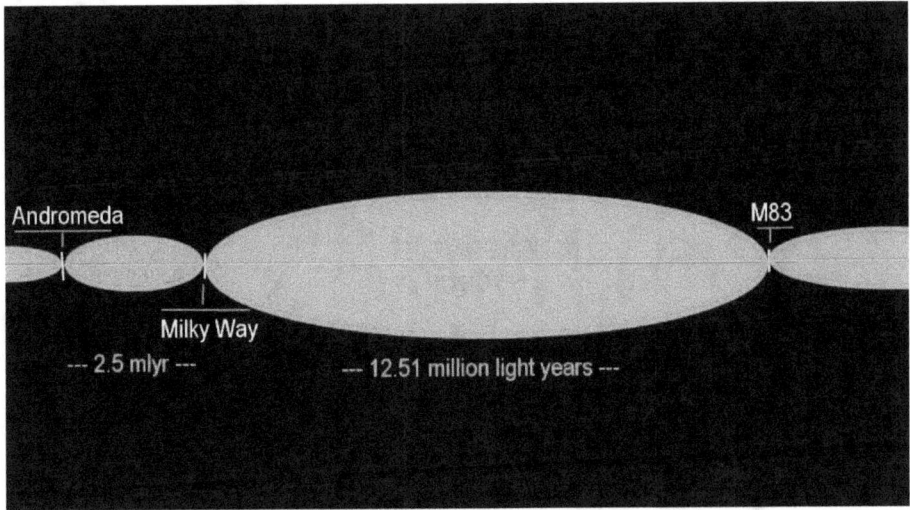

The different resonances would reflect the different distances of the interconnecting plasma streams between the galaxies. The weaker, shorter resonance would be overlaid at the junction at our galaxy onto the larger long cycle. Both, in combination, affect the plasma density in the galaxy, and by reflection, also affect the climate on Earth.

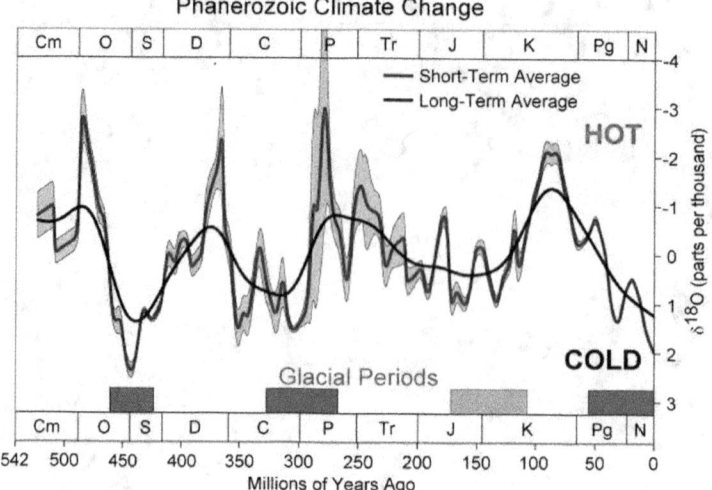

That's what we see reflected in the Phanerozoic climate history, that in turn is likely reflecting the galactic plasma-density history. It is critical to note in this context that the long climate cycles are both near their minimal point. At their combined low levels, the modern Ice Age Epoch began 2 million years ago, termed the Pleistocene Epoch. With our galaxy being thereby at its weakest state in over 400 million years, and with it still getting weaker, it is not surprising that one finds 300 billion stars in the galaxy to have gone inactive already.

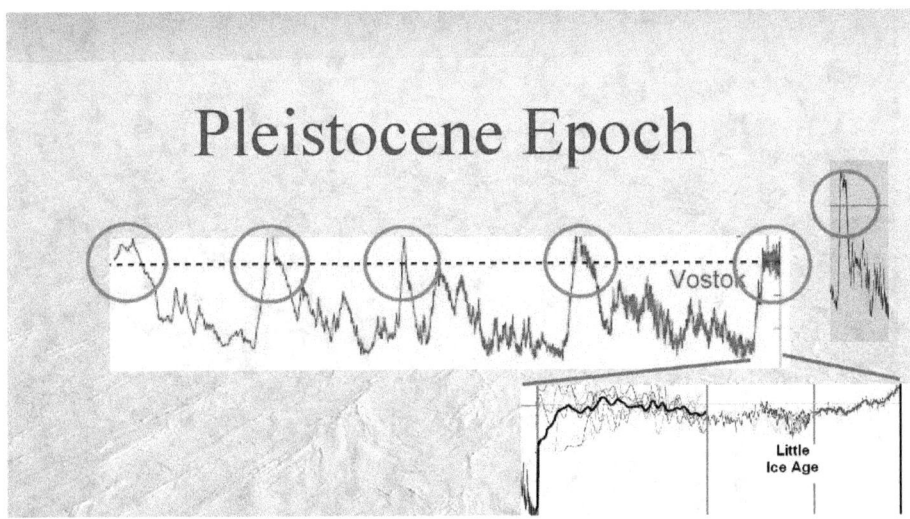

In a few million years into the future, or tens of millions of years, as the case may be, when the galaxy has recovered, the presently inactive stars will then become active again and may remain so for another half a billion years till the next weak point is encountered for the galaxy.

For now, however, the important aspect is the short-term timing. We need to know with a high degree of certainty when the coming phase shift starts as our Sun joins the rank of the 300 billion inactive stars in the galaxy.

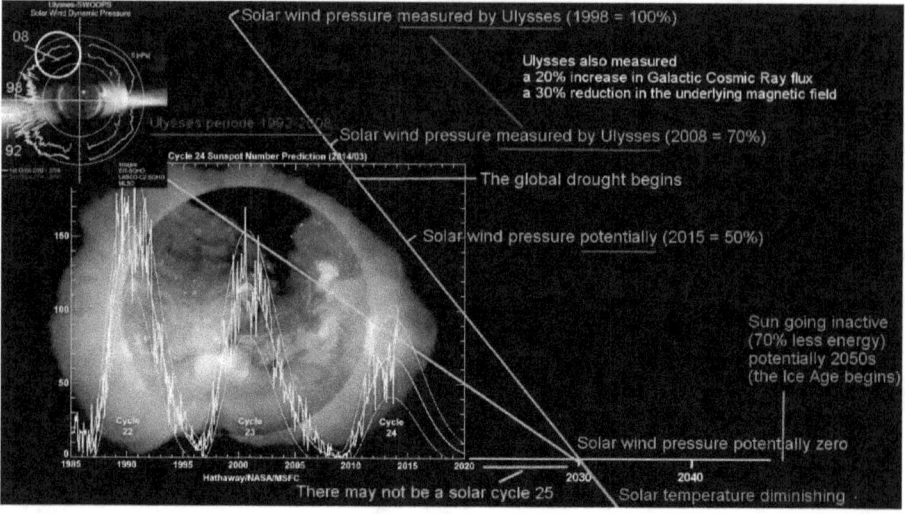

As I said before, the phase shift will likely occur in the 2050s timeframe if not sooner. The solar-activity cycles are already fast diminishing. The solar-wind pressure is fast fading.

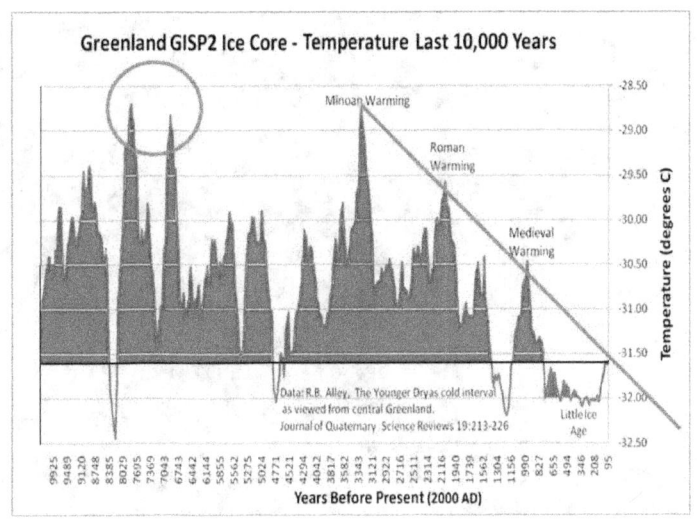

The climate on Earth has also been on a near-term down ramp that began 3000 years ago, which had trailed out into the Little Ice Age, from which the Sun briefly recovered with, potentially, a Dansgaard Oeschger pulse. But with the pulse now ending, the final end of the active Sun is quite near.

If the solar wind continues to diminish

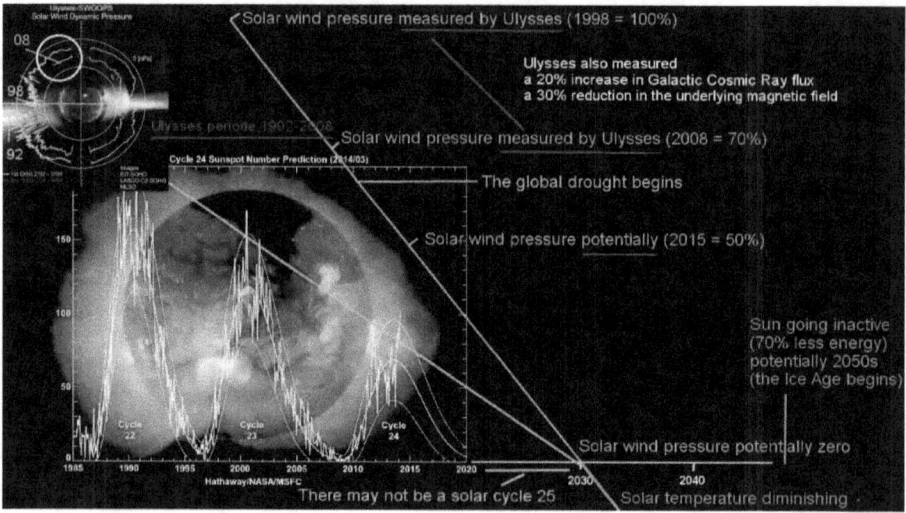

If the solar wind continues to diminish in the short-term, at the rate
that NASA's Ulysses spacecraft has measured, the solar wind will
cease completely in the 2030s. After that the Sun itself will diminish
till its innermost primer fields will collapse. At this point it will
become inactive and revert to the default level in plasma-fusion
intensity, and become a 4,000 K star. This may happen, potentially
in the 2050s, if not sooner.

The timing is important, because it determines the time we have
left for raising up the value of our precious humanity, in society's
self-perception, in order that we may protect our existence before
the most precious asset that we can possibly have on this planet,
which is ourselves, becomes irreversibly thrown away as trash by
neglect. Society is suffering a huge deficit in the arena of its self-
perception.

Saturn eclipsing the Sun

Our humanity is not a trivial gem in the department of life on this planet. We are, by all counts, the greatest gem that has ever lived here. We are the greatest asset on the Earth that we can possibly have. We stand with a history of over two million years of development to our credit, which we rarely give ourselves credit for.

We stand with remarkable achievements in our pocket

Philharmonic Orchestra of Jalisco
(Guadalajara, Jalisco, Mexico)

wikipedia

We stand with remarkable achievements in our pocket, wrought in music, art, culture, science, technology, and humanity, each of which qualify to be termed amazing, extraordinary, if not miraculous.

Main human species

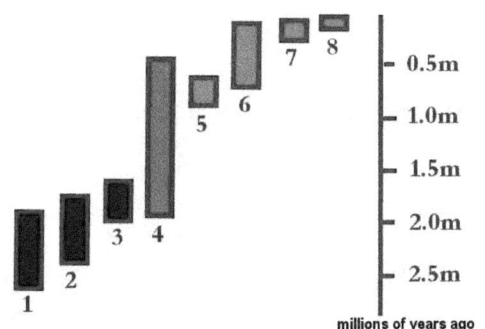

australopithecus rudolfensis (1),
australopithecus habilus (2),
homo ergaster (3),
homo erectus (4),
homo antecessor (5),
homo heidelbergensis (6),
homo neandertalensis (7).
homo sapiens (8)

- 0.5m
- 1.0m
- 1.5m
- 2.0m
- 2.5m

millions of years ago

We, the homo sapiens (8), are the only surviving,
and the shortest lived of all the the human species,
at barely 200,000 years of age.

In the long sweep of the history of humanity, we are presently the last living species in the long train of human development. We named ourselves the homosapien. We are the pinnacle on this course, but are also a child in the larger context. We have a mere 200,000 years to our credit. We are the 8th human species, of which the 7 previous species have all become extinct over time, perhaps in one of the harsh ice ages in the Earth's recent past. Even the most successful species of man, named Homo Erectus, who had lived for more than a million years, has become lost and exists no more.

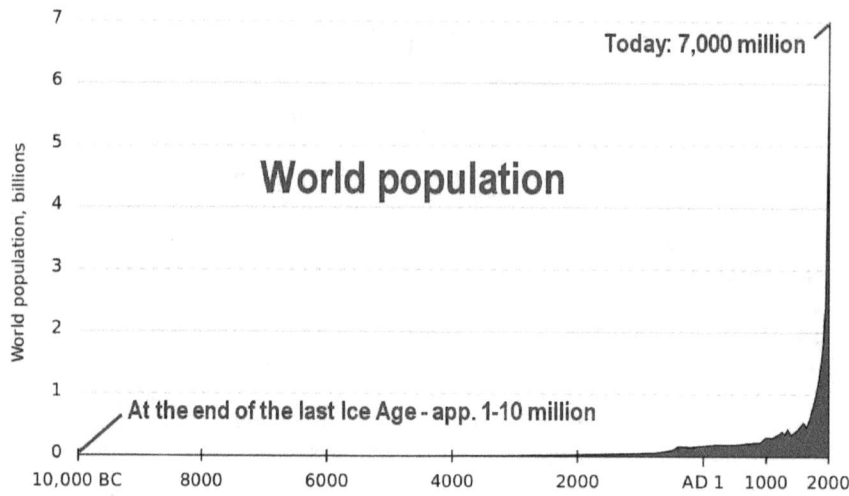

We ourselves, after 200,000 years of development, have emerged from the last Ice Age as the sole survivor with a world population of a mere 1 to 10 million people. This sparse human population on the Earth after the Ice Age, reflects to some degree the harsh conditions in ice age living. Today, we have a world population of seven billion people with a richer culture than could have been imagined only 10,000 years ago; a culture with written languages, great music, astonishing art, architecture, literature, science, technologies, and the freedom to travel to any place on the planet in half a day or less. Human living has become an incredible marvel with an incredible experience that had never before been possible. We have succeeded, beyond anything that had ever been imagined as possible.

We can now see the universe that our world is a part of

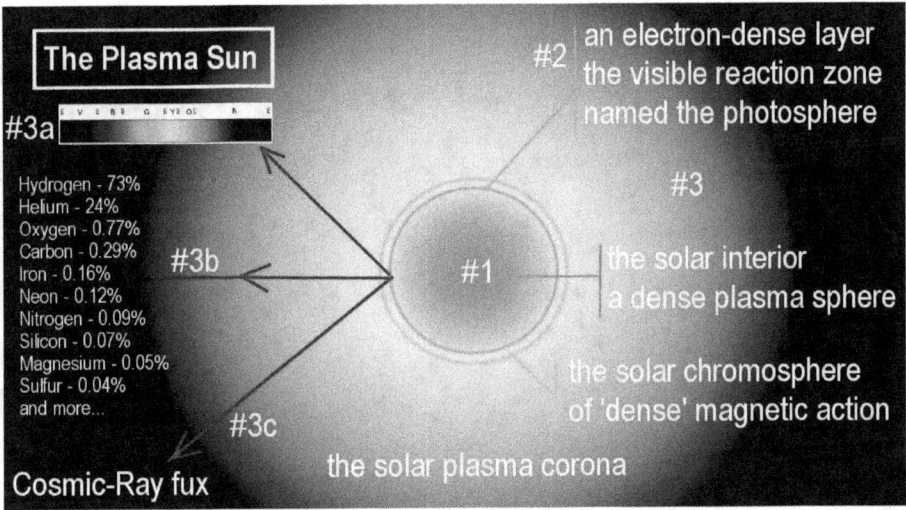

We can now see the universe that our world is a part of. We can even see beyond the leading edge where the mainstream cosmology gets bogged down with epicycles and doctrinal concepts.

We can look at our galaxy and its dynamics

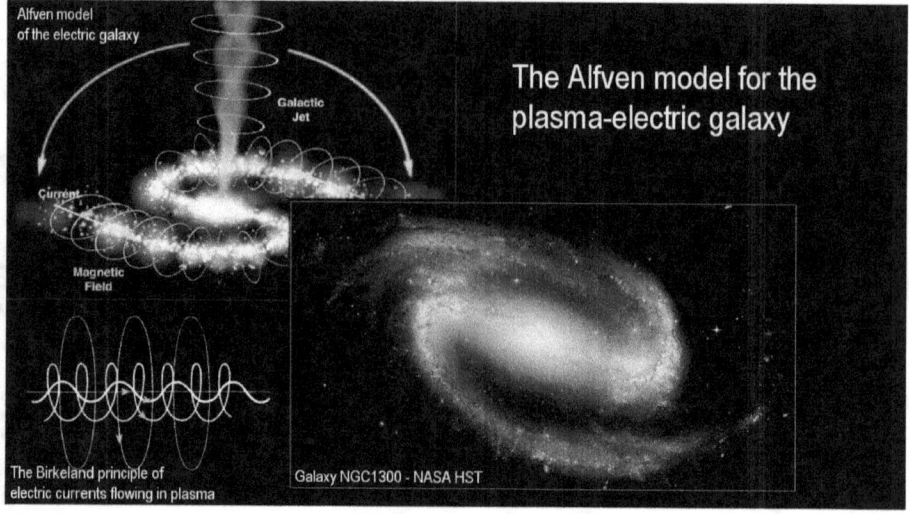

We can look at our galaxy and its dynamics unimpeded, into realms that no one has dared to touch for a long time.

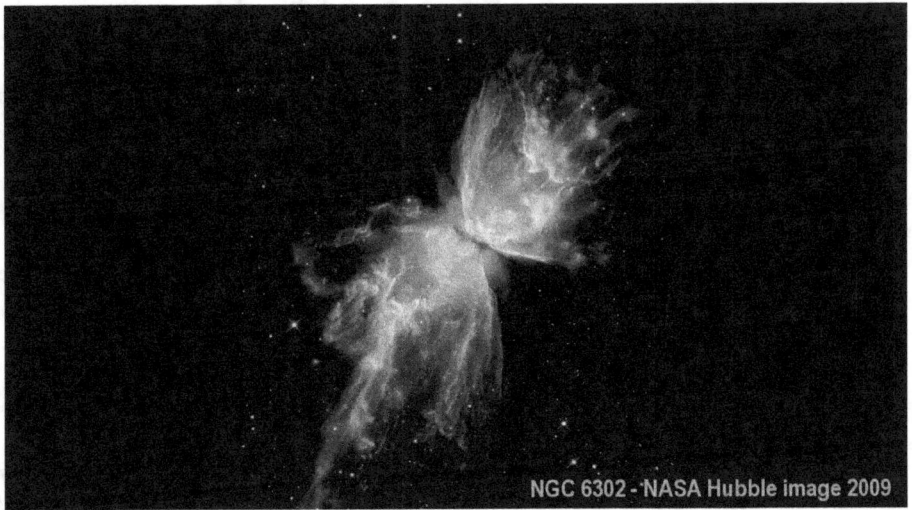

NGC 6302 - NASA Hubble image 2009

By looking at the dynamics that are being discovered with the eye of the mind, we can determine the future that we will face, before it happens.

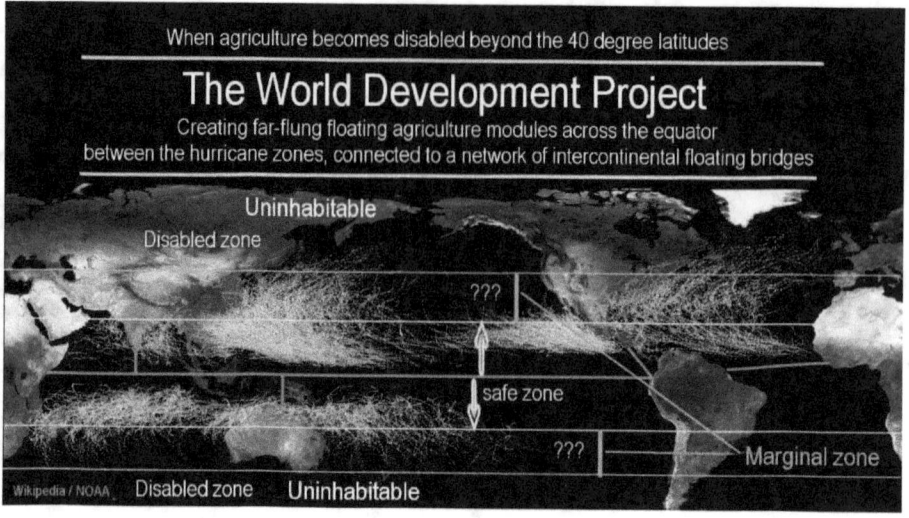

With the knowledge derived thereby, we can shape our world to shape the future before it happens, as we wish to experience it. Of course, what I propose as being essential, which indeed the evidence tells us is existentially critical, is not practical in the general sense of practicality, because the general sense of practicality has become hopelessly too small for anything to be accomplished.

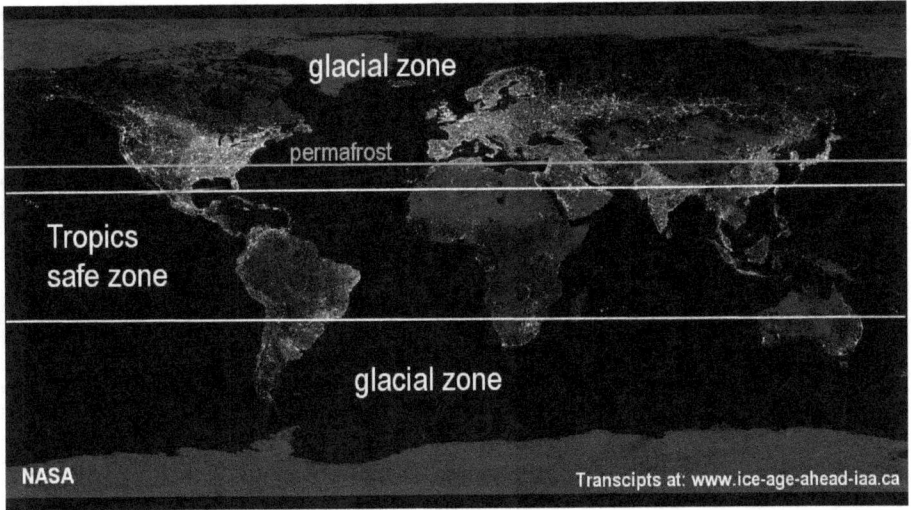

What I propose will most certainly be built, in-spite of it presently being deemed impractical. It will be built, because it is beautiful. The greatness of humanity is beautiful. Great works of art, of science, of infrastructures are beautiful, because they touch a chord within where beauty is anchored. Beauty is the language of the heart; truth is the language of the mind; and love is the language of the soul. With these unfolding as a symphony, civilization will flourish. That's what it means living extraordinarily.

No other expression of life can match the grand capability that we have for the extraordinary, built on small beginnings perhaps, but open to the stars though we may never really fully utilize our potential as it increases even while we move ahead.

Apollo 17 Saturn V rocket on Pad 39-A at dusk.
NASA   This was the last human flight to the moon (Dec. 1972)

Here, in our power to develop ourselves, lies the final frontier. We have contained in us such an amazing heritage, that the gem that we have become, is worthy of the greatest protection, and the greatest nurturing and the most extensive development that is possible.

Will we provide ourselves a future?

But will we do it? Will we provide ourselves a future with the capacity we have as human beings?

Annihilation is assured

500,000 times
Hiroshima
in one hour

Castle Bravo - the first U.S. test of a dry fuel thermonuclear hydrogen bomb - March 1, 1954 at Bikini Atoll, Marshall Islands

Ironically, we are prepared at the present time to throw it all away in a moment of rage. With 500,000 times the destructive force of the Hiroshima bomb, all set up and ready to go, and with the war machine meticulously prepared, ready to become active in the timeframe of a lunch break, humanity seems to have indicated to itself at the present stage that human life. after all, isn't worth living it, or else the nuclear war threats would have ceased long ago.

Instead of removing the danger, the nuclear-war threat has become evermore immanent. But all this is artificial; a tragic small-minded practicality. And being artificial, it can be overturned with the beautiful.

The BRICS leaders in 2014

agenciabrasil.ebc.com.br
at the 2014 G-20 summit
in Brisbane, Australia
by Roberto Stuckert Filho
CC BY 3.0 br via Commons

Putin, Modi, Rousseff, Xi and Zuma

Great steps have already been taken towards resolving the impasse
of ugliness with dedication to the beautiful and the sublime.
On this path, beginning at the grass roots level, both, the nuclear
war challenge, and the Ice Age Challenge, will become resolved
together, for both require the same line of progress to succeed,
from the heart to the heavens.

Two novels by Rolf A. F. Witzsche

Two novels by Rolf A. F. Witzsche
In this context I had started to write two novels, back in the 1980s, as an exploration of how we can get ourselves out of the political trap that had become terribly treacherous even then, and uplift ourselves to our inherent human potential. The two novels that I had started then and eventually completed, were designed to raise our awareness of human living as the most incredible experience of life, and to raise it up further to it being exceedingly precious, which it is.

Then, even while the work was being completed, as if this was not enough, I had started another novel in the early 1990s that became a series of novels, to explore the principle of universal love with the focus on gaining greater freedoms to love one another as children of a common humanity.

A complex of 12 novels with a single story

The Lodging for the Rose

A series of novels by Rolf A. F. Witzsche - part 2

Glass Barriers

Coffee, Sex, and Biscuits

Endless Horizons

Angels of Sex in Queensland

Sword of Aquarius

Lu Mountain

www.ice-age-ahead-iaa.ca

With a focus as wide as that, the project ended more than a decade later as a complex of 12 novels with a single story threaded through the whole.

My point is, that if we don't make the grade on this line and recognize the great value that we have in one another as children of the one single humanity that exists on our planet, the long-prepared-for nuclear Armageddon will surely unfold one tragic morning and by noon the deed will be done that renders the Earth a lifeless rock in the wake of great ensuing agonies, and renders the miracle that we have developed in ourselves, wasted.

- wikipedia
US Navy 100308-N-9588L-023 Machinist's Mate 3rd Class Juan Valles, left, mans the helm of the Ohio-class guided-missile submarine USS Florida (SSGN 728)

The danger is real that we loose everything. The engines stand ready for the button to be pushed. No one will ask afterwards, what was it all for? Nor is this question being asked today, because no rational answers can be given. And tragically, no headway has been made in 70 years towards getting out of the trap that humanity has set for itself.

The only danger that is not artificial

In the same careless manner is the Ice Age danger being brushed aside in today's world, as if it was irrelevant, which in contrast, is the only danger that is not artificial.

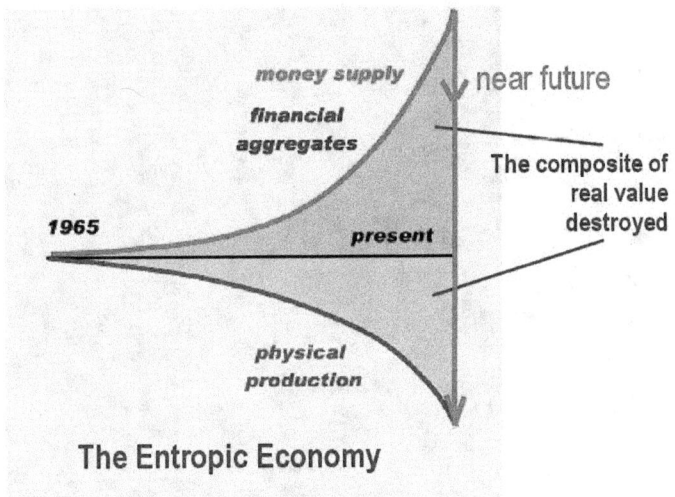

The Entropic Economy

In facing the Ice Age Challenge honestly, which is immensely imperative, we may open up the potential in our heart, and mind, and soul, to heal all the lesser great challenges that have overcome our world already with ever-darker times on the horizon as the world's financial and economic systems break down under the growing weight of speculative looting instead of cooperative creating.

The Ice Age Challenge is not a difficult challenge

The Ice Age Challenge, by itself, is not a difficult challenge to meet, physically. It can be met with floating agriculture and floating cities strung along the equator. However, the footsteps involved are immensely wide in scope, so that they require the cooperative unity of the entire world to master the challenge.

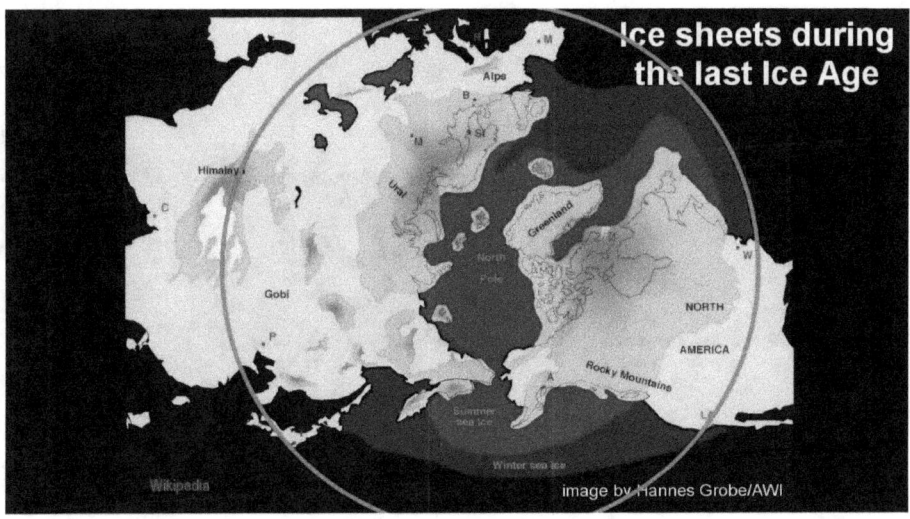

The Ice Age Challenge is always global. No one on this planet is ever not affected by it. It reaches beyond politics, national, regional, or ethnic concerns. The challenge goes to the very center of humanity with the imperative for the world to stand as one for one-another, without exception. We have no hope on any lesser platform, on the Ice Age Front, and by implication on any lesser front.

With our responding to the Ice Age Challenge and its imperatives, with the grandest of all that we have, and are as human beings, we would write ourselves a ticket for building the most sublime and liveable civilization that ever existed, with evermore amazing human experiences along the way as we become one with the universe and one-another.

We stand tall as human beings, on this platform. We stand at the harvest gate. We also know that with each harvest begins the seed time anew.

Seeds are we, wind-blown, by our own wind
Carriers of secrets still unknown
Poems in the words of nature - of our human nature
Sentinels of an Intelligence yet unseen
Prophets of the enduring
Apostles in an endless landscape

Discovering Love
Seeds are we, wind-blown
Poems in the words of nature
Sentinels of an Intelligence yet unseen

Harvest is seedtime, thoughts ripening
Carried as by a great wind
Carriers of secrets to unfold
Thoughts winged with Purpose
A force waiting, silent
Patiently ready for the moment

The Ice Age Challenge
Harvest is seedtime, thoughts ripening
Carried as by a great wind
Carriers of secrets to unfold

Thoughts do awaken
Roused by the moist warmth in spring
Cascades of colors, colors of life
Bright yellows, bursts of silver-white
Thoughts becoming creations
Monuments of genius, builders of worlds

Roses at Dawn in an Ice Age World
Cascades of colors, colors of life
Bright yellows, bursts of silver-white
Thoughts becoming creations

Who owns the seeds? Do we?

Who owns the seeds? Do we?
Who can fathom their wonder?
Life flows from then in great rivers
Rivers trailing into oceans
In them we are alone, each one is alone
Each thought is sovereign, beauty is its song

Who owns the seeds? Do we?
Who can fathom their wonder?
Life flows from them in great rivers
Rivers trailing into oceans
In them we are alone, each one is alone
Each thought is sovereign, beauty is its song

Winning Without Victory
Who owns the seeds? Do we?
Life flows from them in great rivers
Rivers trailing into oceans

Thoughts are seeds, becoming ideas
Alive in discovering
Alive in listening
Alive in being touched by love
Alive in loving
Alive...

Seascapes and Sand
Alive in listening
Alive in being touched by love
Alive in loving

Like seeds, thoughts fall to the ground
Potentials are lost
Hard grounds kill the precious
But we are Man
Hard ground becomes tilled, watered
The precious is nurtured in loving

The Flat Earth Society
Hard grounds kill the precious
But we are Man
Hard ground becomes tilled, watered

Love for one-another, the human spring
Mankind is afloat in a sea that is Love
Seeds germinate, become plants
Roots break the ground
Love lifts the barriers, patiently
Silently waiting, reaching for the sky

Roots break the ground
Love lifts the barriers, patiently
Silently waiting, reaching for the sky

Glass Barriers

Glass Barriers
Roots break the ground
Love lifts the barriers, patiently
Silently waiting, reaching for the sky

Thoughts are the Universe unfolding
Landscapes of brilliance, ideas of power
Substance for enriching one-another
Substance of the forever maturing
Thoughts bearing new seeds within
Seeds for splendours beyond dreams

Coffee, Sex, and Biscuits
Landscapes of brilliance, ideas of power
Substance for enriching one-another
Substance of the forever maturing

Each harvest is seedtime
A seed becomes a plant bearing new seeds
A thought unfolding, bears up civilization
A spark in the heart, bears the 'fire' of life
New worlds are created in the 'fire' of passion
We are the bearers of a 'fire' that is light

Endless Horizons
A thought unfolding, bears up civilization
A spark in the heart, bears the 'fire' of life
New worlds are created in the 'fire' of passion

Builders of worlds are we
New Worlds, which have never been
Precious with riches grander than our own
Nature is Love reflected in loving
Love paints with the colors of its endless spring
Love paints us all - but who owns the seed?

Angels of Sex in Queensland
Precious with riches grander than our own
Nature is Love reflected in loving
Love paints with the colors of its endless spring

Who owns the cradle for the seed?
Name it Intelligence, name it the Universe
Thoughts are seeds from an infinite fountain
Monuments of grandeur of good
Fields of flowers dancing in the sunshine
All nature whispers this to us

Sword of Aquarius
Thoughts are seeds from an infinite fountain
Monuments of grandeur of good
Fields of flowers dancing in the sunshine

The melody of nature - what a song!
Whispers of a splendour grander than the heavens
Like seeds are we - we whisper too
Seeds bearing gifts for the world
Gifts wrapped up in sunshine
Gems are we - unfolding a majestic song!

Lu Mountain

Lu Mountain
Whispers of a splendour grander than the heavens
Like seeds are we - we whisper too
Seeds bearing gifts for the world

Listen to the song

Listen to the song
Listen to the heart
Listen to the silence where strands of love unfold
Listen to the symphony of our humanity
In this symphony we are One
One with the Universe itself.

Flight Without Limits
Listen to the song
Listen to the heart
Listen to the silence where strands of love unfold

Brighter than the Sun
Listen to the symphony of our humanity
In this symphony we are One
One with the Universe itself.